DOCUMENT PROCESSING
A Collegiate Typewriting Simulation

COLLEGIATE PRESS

Contents

Preface

Document Processing is a practical learning simulation designed to provide you with adequate training in the typing of forms that are commonly used in the businesses of today.

The job assignments are introduced in random order to provide a more realistic job situation. You are expected to use the two-letter state abbreviations. To enable you to gain additional experience in the use of other resource materials, we do not provide a list of the state abbreviations.

Many of the job assignments in this book will enable you to use all parts of the typewriter, whether it is electronic, electric, or manual. You will have to use your own judgment in order to make the completed forms mailable.

Mearl Guthrie
Charlene Bunnell

Instructions

MATERIALS

All your forms are provided in this compact workbook so that they cannot be easily damaged or misplaced. Two business forms are provided for each job assignment. The number on the form corresponds to the number of the job assignment. In an effort to keep the cost of this simulation down, some of the forms must be cut to their proper size with scissors.

TYPING THE BUSINESS FORMS

Before beginning your lessons, study the following instructions carefully.

1. *Use current date* on all dated forms. On forms that require past or future dates, appropriate calculations should be made from the current date, unless other instructions are given.

2. *Type data slightly above the printed lines.* Some white space should show between the typewritten line and the printed line. Such letters as "g" and "y" may touch the line but should not overlap it.

3. *Adjust the paper in the machine properly* by use of the variable line spacer and the paper release lever. Use the line finding scale to check the alignment of the printed lines at both the left-hand and right-hand extremes of the form before beginning to type.

4. *Use the space bar* to spell out material to see if it will fit the space provided. Use this procedure if there is any doubt that the space is adequate for the data to be typed. If adequate space is not available, select the best solution from the following possibilities:

    a. *Crowding.* If you have more spaces to type than you have spaces provided, crowd the material to gain as much as two or three spaces.

    b. *Using two lines.* If the material exceeds the space provided by more than two or three spaces, you may find it best to type the data in two lines.

    c. *Abbreviating.* Avoid abbreviations unless it is absolutely necessary or you know that the abbreviation is consistent with accepted procedure. Sometimes the only way to place all the data required in a given space is to abbreviate.

JOB 1
Special Discount Notification

- -

WHAT YOU SHOULD KNOW

In your firm, when a special discount is permitted on an order, you must notify the Billing Department of the details. For such purposes, you use a printed form, Special Discount Notification. This Job consists of preparing two of these notifications. (Note: In typing business forms, usually you may use symbols, such as %.)

a. General Geometric Company, 15 East Weldon Avenue, Phoenix, Arizona 85012-0814, is to be permitted a special discount of 10% on its order of May 21. Total amount of the order is $800. Assume that today's date is May 25.

b. Mercier Brothers, Inc., 196 Fairmount Avenue, Newark, New Jersey 07103-0642, is to be permitted a special chain discount of 10% and 5%. (To compute a chain discount, compute and deduct the first discount; then compute and deduct the second discount. Do not add the two percents together to compute the discount.) Dates are the same as those in the above problem. Total of the order is $1,570.

WHAT YOU ARE TO DO

Fill in the two Special Discount Notifications. Figure the discounts and supply the "Amount Due." Fill in all other necessary information. Sign your name in ink on the line above "Sales Department."

1

<u>SPECIAL DISCOUNT NOTIFICATION</u>

TO: BILLING DEPARTMENT   DATE: _________________

A special discount of ________________ is to be permitted to:

NAME: ____

ADDRESS: ____

               ____

for their order of ____

TOTAL: ____

AMOUNT DUE: ____

SIGNED: ________________________________
               Sales Department

- -

<u>SPECIAL DISCOUNT NOTIFICATION</u>

TO: BILLING DEPARTMENT   DATE: _________________

A special discount of ________________ is to be permitted to:

NAME: ____

ADDRESS: ____

               ____

for their order of ____

TOTAL: ____

AMOUNT DUE: ____

SIGNED: ________________________________
               Sales Department

<u>SPECIAL DISCOUNT NOTIFICATION</u>

TO: BILLING DEPARTMENT   DATE: ___________________

A special discount of _________________ is to be permitted to:

NAME: ____

ADDRESS: ____

                 ____

for their order of ____

TOTAL: ____

AMOUNT DUE: ____

SIGNED: _____________________________
                 Sales Department

- -

<u>SPECIAL DISCOUNT NOTIFICATION</u>

TO: BILLING DEPARTMENT   DATE: ___________________

A special discount of _______________ is to be permitted to:

NAME: ____

ADDRESS: ____

                 ____

for their order of ____

TOTAL: ____

AMOUNT DUE: ____

SIGNED: _____________________________
                 Sales Department

- -

WHAT YOU SHOULD KNOW

Sold to St. John High School, 6030 N. Clifton Avenue, Chicago, Illinois 60640-0182. Shipped to same address. Your order #516.

DESCRIPTION	QUANTITY ORDERED	MON.	TUES.	WED.	THURS.	FRI.	TOTAL
Toothbrushes	250	100	60	50		25	
Toothpaste	250		150		30	70	
Dental Floss	75	10	24	12	8	20	
Dental Powder	2 gross	150	35	15	60	18	

WHAT YOU ARE TO DO

Prepare the Acknowledgment. Type subheadings under "Quantity Received." Make extensions. Compare totals with "Quantity Ordered." At the bottom of the form, type the list of discrepancies you have found.

Acknowledgment of Order

SOLD TO

SHIPPED TO

Your Order No.

	Description	Quantity Ordered	Quantity Received					

NAME: _________________________   JOB 2

<table>
<tr><td colspan="4">Acknowledgment of Order</td></tr>
<tr><td>SOLD TO</td><td colspan="3">SHIPPED TO</td></tr>
</table>

Your Order No.

	Description	Quantity Ordered	Quantity Received					

Transfer of Machines on Maintenance

WHAT YOU SHOULD KNOW

Today Ms. Vera Ross is informing Mr. Henry Johns, principal, about the transfer of some office machines from the Business Education Department of Grant High School, 6000 Overbrook Avenue, Cleveland, Ohio 44107-0547, to the Business Education Department in West High School, 2088 Brookpark Road, Cleveland, Ohio 44109-0663, to be effective the first of next month. The three machines being transferred are:

TYPE AND MODEL	SERIAL NUMBER
IBM Electric	1272615
Royal Electric	7189933
Marchant Calculator	5955487

The transfer will be effective the following month.

WHAT YOU ARE TO DO

Complete the form. Use the current date. Sign your name.

Date:

   To:

From:

Subject: Transfer of Machines on Maintenance

The machines designated below have been transferred from:

    Customer Name   ____
    and Address
                         ____

                         ____

                         ____

This transfer covers the cancellation of the following machines:

    <u>Type and Model</u>   <u>Serial Number</u>   <u>Effective Month</u>

    ________________   ________________   ____________________

    ________________   ________________   ____________________

    ________________   ________________   ____________________

    ________________   ________________   ____________________

    ________________   ________________   ____________________

    ________________   ________________   ____________________

    ________________   ________________   ____________________

Please issue an order covering the installation of these machines at the
following account effective as of the dates shown above:

    Customer Name   ____
    and Address
                         ____

                         ____

                         ____

                       Signed____

Date:

   To:

From:

Subject: Transfer of Machines on Maintenance

The machines designated below have been transferred from:

Customer Name
and Address
      ____

      ____

      ____

      ____

This transfer covers the cancellation of the following machines:

Type and Model	Serial Number	Effective Month
_____________	_____________	_______________
_____________	_____________	_______________
_____________	_____________	_______________
_____________	_____________	_______________
_____________	_____________	_______________
_____________	_____________	_______________
_____________	_____________	_______________

Please issue an order covering the installation of these machines at the
following account effective as of the dates shown above:

Customer Name
and Address
      ____

      ____

      ____

      ____

      Signed_______________________________________

JOB 4
Check Request

- -

WHAT YOU SHOULD KNOW

Mr. Ronald Burns, 814 Elm Avenue, Maumee, Ohio 43537-0357, has been temporarily employed as gardener for a period of 4 days, 8 hours each day, at an hourly rate of $4.25. The check will be written in the Corporate Accounting Department, Room No. 602. The Maintenance Department should be charged for this check.

WHAT YOU ARE TO DO

Use the current date and fill in all blanks that would usually be typed. (The items on the lower part of the form will be filled in by a clerk after the check is written.)

CHECK REQUEST

	Corporate Acctg. Dept.	Room No.
	Domestic Prod. Acctg. Dept.	Room No.
	Marketing Acctg. Dept.	Room No.

DATE _______________________ 19 ___

PLEASE HAVE CHECK ISSUED

TO ORDER OF ____ $ ________________

ADDRESS ____

FOR ____

CHARGE ____

VENDOR			LEG.	DEPT	DIV.	M.C.	G.S.	SEC.	EMER
	V S	E D							

DEPT			DIV. DIV. ST.-DIST.	G. C. POOL ZONE	MGR. DIST. UNIT	SECT. FM'N DIST.	UNIT P. I. ZONE	FARM CAR NO.	WELL	C. G. C. G. C. G.	C. L. C. L. C. L.	L. C. L. C. D. G.	D. G. D. G. D. F.	D. F. D. F. D. I.	D. I. D. I.	QUANTITY OR SUSPENSE NO.		AMOUNT	
X	V S	E D																	
X	V S	E D																	
X	V S	E D																	

NAME: _______________________   JOB 4   21

CHECK REQUEST

Corporate Acctg. Dept.	Room No.	
Domestic Prod. Acctg. Dept.	Room No.	
Marketing Acctg. Dept.	Room No.	

DATE _________________________ 19 ____

PLEASE HAVE CHECK ISSUED _________________

TO ORDER OF ____ $ ________________

____

ADDRESS ____

FOR ____

____

____

____

CHARGE ____

VENDOR			LEG.	DEPT	DIV.	M.C.	G.S.	SEC.	EMER
	V S	E D							

D E P T			DIV. / DIV. / ST.-DIST.	G. C. / POOL / ZONE	MGR. / DIST. / UNIT	SECT. / FM'N / DIST.	UNIT / P. I. / ZONE	FARM / CAR NO.	WELL	C. G. / C. G. / C. G.	C. L. / C. L. / C. L.	L. C. / L. C. / D. G.	D. G. / D. G. / D. F.	D. F. / D. F. / D. I.	D. I. / D. I.	QUANTITY OR SUSPENSE NO.		AMOUNT	
X	V S	E D																	
X	V S	E D																	
X	V S	E D																	

Request for Quotation

WHAT YOU SHOULD KNOW

Request a quotation for the following materials from Toledo Stationers, 29778 Main Street, Toledo, Ohio 43605-0107.

1 gross	3M11 Mimeograph stencils
5 reams	I-16D 8½ × 11 in. Artesian Bond White paper
5 reams	X401 8½ × 11 in. "copy" paper
14 dozen	M241 4⅛ × 9½ in. legal-size envelopes

Quotations will close two weeks from today.

WHAT YOU ARE TO DO

Complete the form, using the current date. Supply the "Quotations Close" date. (The price, terms, and date of shipment will be supplied by the company making the quotations.)

Note: Type and initial your name on the line below "Purchasing Department."

THIS IS A REQUEST FOR A QUOTATION AND <u>NOT AN ORDER</u>

Date...19 ....

Quotations Close:

Date...19 ....

Sirs: Please quote us your lowest price on the Material specified below. If you cannot quote, advise us immediately.

Purchasing Department

By..

QUANTITY	MATERIAL AND DESCRIPTION	PRICE

TERMS	DATE OF SHIPMENT

IMPORTANT

Date of Shipment Must Be Specified

By..

Signed..

NAME: _________________________________   JOB 5   27

THIS IS A REQUEST FOR A QUOTATION AND <u>NOT AN ORDER</u>

Date..**19** ......

Quotations Close:

Date..**19** ......

Sirs: Please quote us your lowest price on the Material specified below. If you cannot quote, advise us immediately.

Purchasing Department

By..

QUANTITY	MATERIAL AND DESCRIPTION	PRICE

TERMS	DATE OF SHIPMENT

IMPORTANT

Date of Shipment Must Be Specified

By..

Signed ....

Additions to the Mailing List

WHAT YOU SHOULD KNOW

One of your jobs is to notify the Mailing Department that new names should be added to the various mailing lists used in your firm. Each mailing list is referred to as a <u>stencil</u>. In each case, you fill in all the necessary information except the last line. It will be filled in by your supervisor who must approve each new list of names. You should always abbreviate the words <u>Avenue</u>, <u>Boulevard</u>, and <u>Street</u>. Also, you should use the official Post Office abbreviations for the names of states. In street names, abbreviate the words <u>North</u>, <u>South</u>, <u>East</u>, and <u>West</u>. No other abbreviations are permitted unless they are abbreviations that a company regularly uses in its title.

WHAT YOU ARE TO DO

Fill in the Additions to Mailing List form for names for Stencil #B-3. Before doing so, arrange the names in strict alphabetical order; then type them in that order. Type "X" to answer the question "No" in the second line from the bottom. Be sure to type your name on the "FROM" line at the top of the form.

C. B. Company, 123 Westcreek Street, Austin, Texas 78749-0727

Catskill Clothiers, 30 West 42 Street, New York, New York 10036-0804

Catskill Clothiers, 28 East 59 Street, New York, New York 10017-0742

Chicago Board of Education, 228 La Salle Street, Chicago, Illinois 60601-0181

C. B. Company, 5 First Avenue, Tallahassee, Florida 32303-0608

C. Robert Jones, 11 Welch Place, Los Angeles, California 90027-0720

Robert C. Jones, 118 Old Gate Road, Lexington, Kentucky 40504-0342

Jones and Samson, Inc., 5 Tulane Drive, Columbus, Ohio 43202-0446

(Continued on following page)

Chicago Bread Co., 228 Adams Street, Chicago, Illinois
60607-0236

Chic Gowns by Alma, 401 Main Street, St. Louis, Missouri
63136-0499

Jonesville Garage, 1043 Hampton Avenue, Hartford,
Connecticut 06110-0213

Lucy Kalmer, 12 Julian Road, Dubuque, Iowa 52001-0143

ADDITIONS TO THE MAILING LIST

TO: MAILING DEPARTMENT FROM: ___________________________ DEPT: ___________________________ DATE: __________

Please add the following names and addresses to Stencil No.________.

	NAME	STREET	CITY	STATE	ZIP
1.					
2.					
3.					
4.					
5.					
6.					
7.					
8.					
9.					
10.					
11.					
12.					
13.					
14.					
15.					

NOTE: If you have additional names, use another form and staple it to this one.

Do you want a proof run-off of these new names? YES _____   NO _____   DATE ___________________________

APPROVED BY ___________________________

ADDITIONS TO THE MAILING LIST

TO: MAILING DEPARTMENT FROM: ______________________________ DEPT: ______________________________ DATE: __________

Please add the following names and addresses to Stencil No.__________.

	NAME	STREET	CITY	STATE	ZIP
1.					
2.					
3.					
4.					
5.					
6.					
7.					
8.					
9.					
10.					
11.					
12.					
13.					
14.					
15.					

NOTE: If you have additional names, use another form and staple it to this one.

Do you want a proof run-off of these new names? YES ______ NO ______ DATE ______________________________

APPROVED BY ______________________________

Service Department Memorandum

- -

WHAT YOU SHOULD KNOW

A typewriter repairperson has just serviced some typewriters in Room 124. Having found some older models and some in need of replacement, she informs her company's Sales Manager so that a company representative may call on the customer (Board of Education) to make suggestions for updating the equipment. Her suggestions include the following:

- New machines needed: 5 Lanier Electronic Typewriters
- Machines to be traded in: 1 Underwood Noiseless, #6699825
- Additional machines advisable: 2 Royal Electric
- Machines in poor condition: Royal Manual, #6671822
- Machines out of use: Remington Model 16, #PR-20935

Ignore the remaining lines.

WHAT YOU ARE TO DO

Complete the form, using the current date. The sales representative will want to contact Ms. Robyn Wood, Room 124, Department of Business, Board of Education, Louisville, Kentucky 40202-0983.

Note: Repairperson's name is Shannon Marie.

SERVICE DEPARTMENT MEMORANDUM

TO SALES MANAGER

DATE_______________________

CUSTOMER'S NAME_______________________

ADDRESS_______________________

ROOM NO._______________________

MR._______________________DEPT._______________________

COMMENTS

NEW MACHINES NEEDED
   (TYPE AND QUANTITY)_______________________

MACHINES SHOULD BE TRADED-IN
   (TYPE AND SERIAL NUMBER)_______________________

ADDITIONAL MACHINES ADVISABLE
   (TYPE AND QUANTITY)_______________________

MACHINES IN POOR CONDITION
   (TYPE AND SERIAL NUMBER)_______________________

MACHINES OUT OF USE
   (TYPE AND SERIAL NUMBER)_______________________

CYLINDER INFORMATION_______________________

ACCESSORIES NEEDED_______________________

MIXED EQUIPMENT_______________________

PURCHASES OF SECOND-HAND EQUIPMENT_______________________

IS THIS A MACHINE SALE TIP?_______________________

INSTRUCTION IN USE OF MACHINES ADVISABLE_______________________

REMARKS_______________________

____

____

____

____

____
SERVICEMAN

SERVICE DEPARTMENT MEMORANDUM

TO SALES MANAGER

DATE________________________

CUSTOMER'S NAME____

ADDRESS____

ROOM NO.____

MR.________________________________DEPT.________________

COMMENTS

NEW MACHINES NEEDED
(TYPE AND QUANTITY)____

MACHINES SHOULD BE TRADED-IN
(TYPE AND SERIAL NUMBER)__________________________________

ADDITIONAL MACHINES ADVISABLE
(TYPE AND QUANTITY)____

MACHINES IN POOR CONDITION
(TYPE AND SERIAL NUMBER)__________________________________

MACHINES OUT OF USE
(TYPE AND SERIAL NUMBER)__________________________________

CYLINDER INFORMATION.____________________________________

ACCESSORIES NEEDED______________________________________

MIXED EQUIPMENT____

PURCHASES OF SECOND-HAND EQUIPMENT______________________

IS THIS A MACHINE SALE TIP?________________________________

INSTRUCTION IN USE OF MACHINES ADVISABLE__________________

REMARKS____

____

____

____

____

____

____
SERVICEMAN

Dental Assistant Application

WHAT YOU SHOULD KNOW

Judy Murray wishes to apply for a position as dental assistant. Her personal data follow:

- Address: 7654 Bay Shore, Long Beach, California 90803-0021.
- Telephone: (213) 353-0893.
- Experience: Typing, Shorthand, assisting dentist at chair.
- Education: 2 years, Hillcrest College.
- Salary Expected: Negotiable.
- Position Desired: Full-time.
- Former Employer: James Olson, D.D.S., Long Beach, California, as a Dental Assistant (Telephone: (213) 475-4869).

WHAT YOU ARE TO DO

Fill in all necessary information. Use the current date.

DENTAL ASSISTANT APPLICATION

Name.. Date..

Street... Telephone...

City.. State...

Experience: ☐ Typing   ☐ Shorthand   ☐ Bookkeeping
                             ☐ Assisting doctor at chair   ☐ Laboratory work

Education...

Salary expected $.............................. Per............................

Position desired: ☐ Full Time   ☐ Part Time   ☐ Temporary

Former Employer   Address   Position

....

....

         Reference   Address   Telephone

....

....

Remarks

DENTAL ASSISTANT APPLICATION

Name..   Date..

Street..   Telephone..

City...   State..   ....

Experience: ☐ Typing   ☐ Shorthand   ☐ Bookkeeping
☐ Assisting doctor at chair   ☐ Laboratory work

Education..

Salary expected $.................................   Per...

Position desired: ☐ Full Time   ☐ Part Time   ☐ Temporary

Former Employer	Address	Position
....	....	....
....	....	....

Reference	Address	Telephone
....	....	....
....	....	....

Remarks

JOB 9
Expediting Order

- -

WHAT YOU SHOULD KNOW

Your department has not received shipment of stationery supplies that were ordered. These supplies are needed, and you need to learn when shipment will be made. Use the current date. Purchase Order No. 4839, dated one month previous. Destination is King High School, Business Education Department. Date material required is one week from today. Order reference 4869, six weeks previous. Date promised was three work days previous.

WHAT YOU ARE TO DO

Use material listed in Job #5—Toledo Stationers.

<u>THIS IS NOT AN ORDER</u>

Date:...

Our Purchase
Order No. .... Dated ....

Destination ....

Date Material Required ....

Your Order
Reference .... Dated ....

Date Promised ....

We are EXPEDITING this purchase order.   Please complete this form and return it to us promptly for our records.

Material:

.... When will shipment be made? ....

.... Was shipment made as promised? .... If shipped please forward invoice in triplicate with shipping papers attached.

.... If not shipped, give revised shipping date ....

.... Please forward invoice in triplicate with shipping papers attached for shipment.

Received
Made   on ....

Remarks:

....
PURCHASING AGENT

By: ....

THIS IS NOT AN ORDER

Date: ....

Our Purchase
Order No. .... Dated ....

Destination ....

Date Material Required ....

Your Order
Reference .... Dated ....

Date Promised ....

We are EXPEDITING this purchase order. Please complete this form and return it to us promptly for our records.

Material:

.... When will shipment be made? ....

.... Was shipment made as promised? .... If shipped please forward invoice in <u>triplicate</u> with shipping papers attached.

.... If not shipped, give revised shipping date ....

.... Please forward invoice in <u>triplicate</u> with shipping papers attached for shipment.

Received
Made   on ....

Remarks:

....
PURCHASING AGENT

By: ....

Receiving and Inspection Report

- -

WHAT YOU SHOULD KNOW

The following material was purchased from the Morris Manufacturing Company, 1964 Shannon Avenue, Milwaukee, Wisconsin 53207-0708.

ITEM	QUANTITY	STOCK NO.	VENDOR'S NO. & DESCRIPTION		NO.	WEIGHT
1	500	5609	7401	Spark plug housings	5	1,000
2	5000	5739	9308	Generator brushes	50	10,000
3	500	9379	2231	Generator housings	5	1,000
4	5000	8422	ES1716	Electric starter cases	50	10,000
5	1500	385081	EF1761	Electric motor frames	15	3,000
6	2000	375	9448	Generator shafts	20	4,000
7	5000	624900	2340	Spark plug insulators	50	10,000
8	500	458	49091	Stoplight frames	5	1,000

The frames are shipped in bundles, the housings in barrels, the brushes in cartons, the insulators in boxes, the shafts in crates, and the cases in packages. All items come 100 units in a container. Terms are 2/10, n/30. All items were received today via prepaid rail freight, Pro. No. 6539608C, from the factory in Milwaukee, Wisconsin. Total weight: 40,000 lb. Purchase Order #7095.

WHAT YOU ARE TO DO

Complete the form. Compute the number of containers. All material was received today without apparent shipping damages. Use the current date.

RECEIVING AND INSPECTION REPORT

RECEIVED FROM		RECEIVED BY	DATE RECEIVED

RECEIVED VIA

☐ UPS ☐ VENDOR TRUCK
☐ TRUCK ☐ AIR FRT.
☐ PP ☐ OTHER

PRO. NO.

NUMBER OF CONTAINERS FREIGHT CHARGES:
☐ PREPAID
☐ COLLECT

☐ SHIPMENT COMPLETE ☐ ORDER COMPLETE
☐ PARTIAL SHIPMENT ☐ PARTIAL ORDER

STOCK NUMBER	QUANTITY	UNIT	DESCRIPTION AND VENDOR'S STOCK NO.	ORDER NO.

SHIPPING DAMAGE— DESCRIBE NATURE AND EXTENT OF ANY SHIPPING DAMAGE NOTED:

INSPECTED BY: DATE:

RECEIVING INSPECTION— DESCRIBE NATURE AND EXTENT OF ANY DEFECT NOTED. ATTACH ADDITIONAL SHEET IF NEEDED:

☐ BROKEN ☐ CONTAMINATED
☐ BENT ☐ PACKAGE OR LABELING
☐ FINISH ☐ WET
☐ COMPONENT MISSING ☐ SOILED
☐ NON-FUNCTIONAL ☐ OTHER
☐ QUANTITY DISCREPANCY ☐ **ACCEPTED**

INSPECTED BY: DATE:

INITIAL ACTION TAKEN:

BY: DATE:

FINAL DISPOSITION:

BY: DATE:

RECEIVING AND INSPECTION REPORT

RECEIVED FROM	RECEIVED BY	DATE RECEIVED

RECEIVED VIA

☐ UPS   ☐ VENDOR TRUCK
☐ TRUCK   ☐ AIR FRT.
☐ PP   ☐ OTHER

PRO. NO.

NUMBER OF CONTAINERS	FREIGHT CHARGES:

☐ PREPAID
☐ COLLECT

☐ SHIPMENT COMPLETE   ☐ ORDER COMPLETE
☐ PARTIAL SHIPMENT   ☐ PARTIAL ORDER

STOCK NUMBER	QUANTITY	UNIT	DESCRIPTION AND VENDOR'S STOCK NO.	ORDER NO.

SHIPPING DAMAGE— DESCRIBE NATURE AND EXTENT OF ANY SHIPPING DAMAGE NOTED:

INSPECTED BY: DATE:

RECEIVING INSPECTION— DESCRIBE NATURE AND EXTENT OF ANY DEFECT NOTED. ATTACH ADDITIONAL SHEET IF NEEDED:

☐ BROKEN
☐ BENT
☐ FINISH
☐ COMPONENT MISSING
☐ NON-FUNCTIONAL
☐ QUANTITY DISCREPANCY

☐ CONTAMINATED
☐ PACKAGE OR LABELING
☐ WET
☐ SOILED
☐ OTHER
☐ **ACCEPTED**

INSPECTED BY: DATE:

INITIAL ACTION TAKEN:

BY: DATE:

FINAL DISPOSITION:

BY: DATE:

Bad Check Notice

WHAT YOU SHOULD KNOW

A check for $85.00 issued to the Burr Oak Sports Shop (dated one week ago) was returned by the First National Bank of Detroit to you. The check was written by Mrs. John DeForte, 1300 Oak Street, Detroit, Michigan 48124-2677.

WHAT YOU ARE TO DO

Complete the Bad Check Notice using the current date. The address of the Burr Oak Sports Shop is 623 Eagleton Street, Burr Oak, Michigan 49030-0145. The check was stamped "Account Closed."

TO ☐   **BAD CHECK NOTICE**

THIS IS TO INFORM YOU THAT I AM IN RECEIPT OF A CHECK ALLEGED TO HAVE BEEN WRITTEN BY YOU,

DATED	MADE PAYABLE TO	NAME OF BANK DRAWN ON	AMOUNT

THIS CHECK WAS PRESENTED TO ME IN THE USUAL COURSE OF BUSINESS, AND WAS RETURNED TO ME FROM THE ABOVE SAID BANK MARKED,

☐ INSUFFICIENT FUNDS   ☐ ACCOUNT CLOSED

IN ACCORDANCE WITH THE MICHIGAN STATUTE YOU ARE HEREBY GIVEN FIVE (5) DAYS NOTICE THAT SAID CHECK HAS NOT BEEN PAID, AND IF YOU SHALL NOT HAVE PAID THE AMOUNT DUE THEREON WITHIN FIVE (5) DAYS OF RECEIPT OF THIS NOTICE, THIS SHALL SERVE AS EVIDENCE OF INTENT TO DEFRAUD, AND A REQUEST TO THE OFFICE OF THE PROSECUTING ATTORNEY TO TAKE CRIMINAL ACTION WILL BE MADE BY ME.

SIGNED _______________________________

ADDRESS _______________________________

____

DATED THIS _________________ DAY OF _________________ 19_____

TO [ ]

BAD CHECK
NOTICE

THIS IS TO INFORM YOU THAT I AM IN RECEIPT OF A CHECK ALLEGED TO HAVE BEEN WRITTEN BY YOU,

DATED	MADE PAYABLE TO	NAME OF BANK DRAWN ON	AMOUNT

THIS CHECK WAS PRESENTED TO ME IN THE USUAL COURSE OF BUSINESS, AND WAS RETURNED TO ME FROM THE ABOVE SAID BANK MARKED,

☐ INSUFFICIENT FUNDS   ☐ ACCOUNT CLOSED

IN ACCORDANCE WITH THE MICHIGAN STATUTE YOU ARE HEREBY GIVEN FIVE (5) DAYS NOTICE THAT SAID CHECK HAS NOT BEEN PAID, AND IF YOU SHALL NOT HAVE PAID THE AMOUNT DUE THEREON WITHIN FIVE (5) DAYS OF RECEIPT OF THIS NOTICE, THIS SHALL SERVE AS EVIDENCE OF INTENT TO DEFRAUD, AND A REQUEST TO THE OFFICE OF THE PROSECUTING ATTORNEY TO TAKE CRIMINAL ACTION WILL BE MADE BY ME.

SIGNED _______________________________

ADDRESS _______________________________

____

DATED THIS _______________ DAY OF _______________ 19_______

NAME: _______________________   JOB 11   65

Application for Credit

WHAT YOU SHOULD KNOW

Most businesses do not pay cash to their suppliers for the merchandise they purchase; instead, they arrange for credit with their major suppliers. Before a supplier can grant credit terms to a business, the business must complete an Application for Credit form.

Your boss, Ms. Carlene Down, sole proprietor of the Burr Oak Sports Shop, is applying for credit with the Action Suit Company, 129 Action Lane, Forbesville, Georgia 31561-2428. Ms. Down has owned the shop for six years.

Additional information you will need to complete the form:

Burr Oak Sports Shop
623 Eagleton Street
Burr Oak, Michigan 49030-0145

Ms. Carlene Down
968 Western Road
Burr Oak, Michigan 49030-0190
Telephone: (616) 489-6118

Finance information:

First National Bank of Burr Oak
J. C. Kline, President
101 Main Street
Burr Oak, Michigan 49030-0101

Reference information:

Gateway Sports Supplies
809 Midway Avenue

Sturgis, Michigan 49091-6019
Telephone: (616) 651-4550

Aerobics Equipment
202 S. Queens Avenue
New York, New York 10042-3248
Telephone: (212) 517-3985

Complete the form using the current date. Ms. Down will sign the application when she returns from a buying trip next week.

BY

Name of firm or individual

Address _______________________ ______ No. of years at this address

City, State, Zip Code _______________________ ____________ Telephone

HEREBY apply for credit in accordance with the terms of:

TO

_______________________ Credit Mgr.

_______________________ Our normal credit terms

The following information must be completed in full; and will be held in the strictest confidence.

OWNERSHIP

☐ Corporation   ☐ Partnership   ☐ Individual
☐ Check here if incorporated within the last 12 months

Name(s) of Principal(s)   Address   Phone

FINANCE

Bank   Bank address

Bank officer or department

REFERENCES

Business name   Address   Phone

☐ Check here if cash sales are okay until credit is approved

We certify that all the information on this form is correct; and that we fully understand your credit terms and agree to the proper payment in consideration of extended credit.

(signed) _______________________

Date _______________ 19 ______   (title) _______________________

Please do not write in the space below

References checked by   ☐ Credit approved, by

Reference results   ☐ Credit refused, by

                                 Date

Name of firm or individual

BY

Address _______ No. of years at this address

_______ Telephone

City, State, Zip Code

HEREBY apply for credit in accordance with the terms of:

TO

Credit Mgr.

Our normal credit terms

The following information must be completed in full; and will be held in the strictest confidence.

FOLD

FOLD HERE
TO FIT
WINDOW ENV.

OWNERSHIP

☐ Corporation   ☐ Partnership   ☐ Individual
☐ Check here if incorporated within the last 12 months

Name(s) of Principal(s)   Address   Phone

FINANCE

Bank   Bank address

Bank officer or department

Business name   Address   Phone

REFERENCES

☐ Check here if cash sales are okay until credit is approved

We certify that all the information on this form is correct; and that we fully understand your credit terms and agree to the proper payment in consideration of extended credit.

(signed) _______

Date _______ 19 _______   (title) _______

Please do not write in the space below

References checked by

☐ _______
Credit approved, by

Reference results

☐ _______
Credit refused, by

Date

Office Machine Record Card

WHAT YOU SHOULD KNOW

An Office Machine Record Card is to be made out for an IBM Model C typewriter used in the Sales Department. Office serial number is 1271615. The machine was ordered six months ago, Purchase Order No. 7268, and the price was $1,169, with 4% sales tax added. A service contract with IBM became effective two weeks after purchase date. The rate of this contract is $128 a year, with four inspections. Room number is 306.

WHAT YOU ARE TO DO

Use the current date to determine other dates. The information at the top and bottom of the card will be filled in with a key punch later.

X

Year	Make of Machine	Serial No.	Location	Dept.	Make	Style	Room No.	Serviced By	Amount

Dept. Chg.	Purchase Order		
Year	Purchase Date		Price
Make	Model		Federal Tax
			Sales Tax
	Location		Total
Serial No.	Remarks		
Location			
Dept.			
Make Style	Service Contract Signed	Rate Per Year	
Room No.	Effective	No. Inspections	
Serviced By	Serviced By		
Amount	Final Disposition		

Year	Make of Machine	Serial No.	Location	Dept.	Make	Style	Room No.	Serviced By	Amount

1 2 3 4 5 6 7 8 9 10 11 12 13 14 15 16 17 18 19 20 21 22 23 24 25 26 27 28 29 30 31 32 33 34 35 36 37 38 39 40 41 42 43 44 45 46 47 48 49 50 51 52 53 54 55 56 57 58 59 60 61 62 63 64 65 66 67 68 69 70 71 72 73 74 75 76 77 78 79 80

X

X

	Year	Make of Machine	Serial No.	Location	Dept.	Make	Style	Room No.	Serviced By	Amount

Dept. Chg.		Purchase Order								
Year		Purchase Date						Price		
Make		Model						Federal Tax		
								Sales Tax		
		Location						Total		
Serial No.		Remarks								
Location										
Dept.										
Make	Style	Service Contract	Signed	Rate Per Year						
Room No.			Effective	No. Inspections						
Serviced By				Serviced By						
Amount		Final Disposition								

Year	Make of Machine	Serial No.	Location	Dept.	Make	Style	Room No.	Serviced By	Amount

1 2 3 4 5 6 7 8 9 10 11 12 13 14 15 16 17 18 19 20 21 22 23 24 25 26 27 28 29 30 31 32 33 34 35 36 37 38 39 40 41 42 43 44 45 46 47 48 49 50 51 52 53 54 55 56 57 58 59 60 61 62 63 64 65 66 67 68 69 70 71 72 73 74 75 76 77 78 79 80

Employees Weekly Time Schedule

WHAT YOU SHOULD KNOW

This form is a weekly time schedule kept by the secretary for the entire department or office. It is prepared at the end of the week or on Monday of the following week. The form must be filled out and then distributed for the proper signatures. The employees and their daily hours for last week are:

Lissy C. Baird	8	8	5	S	S	0	0
J. H. Dyess	1	1	1	1	1	0	0
Betty L. Powell	8	8	8	8	8	0	0
Willie W. Shipp	8	8	8	8	8	0	0
M. L. Sneed	1	1	1	1	1	0	0
Shirley F. Whitter	8	8	8	8	8	0	0

The district is East Texas. The time schedule for J. H. Dyess and M. L. Sneed is figured by the day instead of the hour. The department head is L. A. Bell. Lissy C. Baird was ill Wednesday afternoon and did not return to work on Thursday or Friday.

WHAT YOU ARE TO DO

Complete the form from the information provided. Use the current date.

EMPLOYEES WEEKLY TIME SCHEDULE
SALARIED EMPLOYEES

WEEK ENDED ________________________________   DEPARTMENT OR DISTRICT ________________________________

EMPLOYEE	ACTUAL HOURS WORKED								DO NOT FILL IN			EMPLOYEE'S CERTIFICATION AS TO ACCURACY OF HOURS WORKED
	Mon.	Tues.	Wed.	Thur.	Fri.	Sat.	Sun.	TOTAL HOURS	OT	DT	ST	

Report to be signed by each employee listed thereon.

Do not include hours for off-time not actually worked. Report any off-time or absence by using the following symbols.

B - Serious illness in immediate family (state relationship).

D - Death in immediate family (state relationship).

S - Sick   V - Vacation

H - Holiday   J - Jury Duty   A - Other Absence (state reason)

____

DEPT. HEAD

Department Head may make recommendation as to pay for absence not governed by policy.

EMPLOYEES WEEKLY TIME SCHEDULE
SALARIED EMPLOYEES

WEEK ENDED _________________________________   DEPARTMENT OR DISTRICT _______________________________

EMPLOYEE	ACTUAL HOURS WORKED								DO NOT FILL IN			EMPLOYEE'S CERTIFICATION AS TO ACCURACY OF HOURS WORKED
	Mon.	Tues.	Wed.	Thur.	Fri.	Sat.	Sun.	TOTAL HOURS	OT	DT	ST	

Report to be signed by each employee listed thereon.
Do not include hours for off-time not actually worked. Report any off-time or absence by using the following symbols.
B - Serious illness in immediate family (state relationship).
D - Death in immediate family (state relationship).
S - Sick   V - Vacation
H - Holiday   J - Jury Duty   A - Other Absence (state reason)

Department Head may make recommendation as to pay for absence not governed by policy.

____
DEPT. HEAD

NAME: _______________________________   JOB 14   83

Request for Warranty Bond

WHAT YOU SHOULD KNOW

This form is a request by the owner of property on which a new roof has been installed for a bond covering the roof. The bond is a warranty for the material for a set number of years. Joe Mercer, 4358 Maple Avenue, Dallas, Texas 75219-9821, is requesting such a bond covering 24 squares of 235# White Square-Tabs Asphalt Shingles, which were used in completing a reroofing job on a new building located at 1235 Oak Lawn Avenue, Dallas, Texas 75207-5826. The composition roof had been damaged by hail, and the roofing contractor, Norris Sheet Metal and Roofing Company, 1425 Marburg Street, Dallas, Texas 75215-3689, completed the new roof one month ago today. The code number on the Shingle Wrapper is 8-D20-62, and the slope of the roof deck is 5″ in 12″. Value of the roof is $2,385.

WHAT YOU ARE TO DO

Complete the form for Mr. Mercer. Use the current date.

REQUEST FOR ASPHALT SHINGLE ROOF BOND

DATE_________________________

Gentlemen:

We hereby notify you that we have installed, according to your application instructions,_____________________squares of

____Asphalt Shingles manufactured by you on the building
    WEIGHT   COLOR   TYPE

OWNED BY____
      NAME   STREET ADDRESS   CITY   STATE

LOCATED AT____
      STREET ADDRESS   CITY   STATE

CODE NO. ON SHINGLE WRAPPER__________________________________SLOPE OF ROOF DECK__________________

COMPLETION DATE____NEW BUILDING? ☐ YES ☐ NO

If Re-Roofing Job, describe type and condition of old roof over which shingles were applied______________________

____

Value of applied roof (exclusive of flashings and metal work) to be covered by bond $_________________________

Please issue your bond for asphalt shingle roof in accordance with the foregoing.

Name of Roofing Contractor or Applicator____

Signature____

Address____

City______________________________________State_____________________

Place X in square ☐ preceding weight of shingle applied.

☐ 195# 2-TAB HEX	10 Year Bond	☐ 245# 10x40 SQUARE TABS	15 Year Bond
☐ 195# 3-TAB HEX	10 Year Bond	☐ 240# LOK-TABS	15 Year Bond
☐ 180# STANDARD TITE-ON	15 Year Bond	☐ 235# SQUARE TABS	15 Year Bond
☐ 250# DUBL-COVERAGE TITE-ON	17 Year Bond	☐ 235# SELF-SEALING	15 Year Bond
☐ 165# MASS. WGT. CLIPTFAST	15 Year Bond	☐ 265# SUBURBANS	20 Year Bond
☐ 165# ANGLE DUTCH LAP	15 Year Bond	☐ 300# SUBURBANS	25 Year Bond
☐ 165# MASS. WGT. DUTCH LAP	15 Year Bond		

DO NOT WRITE BELOW THIS LINE

Approved_____________________________Date_________________________
         SALES MANAGER

Bond Certificate No.__________________Dated________________Expiration Date_________________

REQUEST FOR ASPHALT SHINGLE ROOF BOND

DATE_______________________

Gentlemen:

We hereby notify you that we have installed, according to your application instructions,__________________squares of

____Asphalt Shingles manufactured by you on the building
    WEIGHT   COLOR   TYPE

OWNED BY____
        NAME   STREET ADDRESS   CITY   STATE

LOCATED AT____
           STREET ADDRESS   CITY   STATE

CODE NO. ON SHINGLE WRAPPER_____________________________SLOPE OF ROOF DECK______________

COMPLETION DATE_______________________________________NEW BUILDING? ☐ YES ☐ NO

If Re-Roofing Job, describe type and condition of old roof over which shingles were applied________________________________

____

Value of applied roof (exclusive of flashings and metal work) to be covered by bond $______________________

Please issue your bond for asphalt shingle roof in accordance with the foregoing.

Name of Roofing Contractor or Applicator____

        Signature____

        Address____

        City_____________________________State___________________

Place X in square ☐ preceding weight of shingle applied.

☐ 195# 2-TAB HEX	10 Year Bond	☐ 245# 10x40 SQUARE TABS 15 Year Bond
☐ 195# 3-TAB HEX	10 Year Bond	☐ 240# LOK-TABS 15 Year Bond
☐ 180# STANDARD TITE-ON	15 Year Bond	☐ 235# SQUARE TABS 15 Year Bond
☐ 250# DUBL-COVERAGE TITE-ON	17 Year Bond	☐ 235# SELF-SEALING 15 Year Bond
☐ 165# MASS. WGT. CLIPTFAST	15 Year Bond	☐ 265# SUBURBANS 20 Year Bond
☐ 165# ANGLE DUTCH LAP	15 Year Bond	☐ 300# SUBURBANS 25 Year Bond
☐ 165# MASS. WGT. DUTCH LAP	15 Year Bond	

DO NOT WRITE BELOW THIS LINE

        Approved___________________________Date___________________
                  SALES MANAGER

Bond Certificate No._________________Dated______________Expiration Date_______________

JOB 16
Installment Loan Application

WHAT YOU SHOULD KNOW

This form is an application for an installment loan from a bank. Such a loan is made to an individual or a company. You will need the following information:

Amount of loan, $6,000; number of months, 30; purpose, European trip; name of applicant, Betty R. Fulton; telephone number, 437-3431; address, 3465 North Polk Street, Dallas, Texas 75208-2367; period lived at this address, 24 years, 6 months; age, 25 years, marital status, single; number of dependents, 2; date of payments preferred, 1st of each month; employer, Ford Motor Co.; address of employer, 1542 E. Illinois Avenue, Dallas, Texas 75216-6531; telephone number of employer, (214) 918-4165; how long there, 4 years, 4 months; position, stenographer; department, sales; kind of business, manufacturer of cars; monthly salary, $1,450; previous employer, Continental Bus Company; address of previous employer, 411 W. Main; how long there, 2 years (part-time); name, address, and relationship of two close relatives, Mrs. B. J. Camp, 4001 W. Story, sister and Mr. J. E. Smith, 4105 North Elk, brother; bank account, First National (Checking and Savings); title of account, Ms. Betty R. Fulton; credit union, Ford Credit; value of shares, $2,200; automobile owned, 19— Ford Mustang; financed by, Ford Credit Association; amount owing, $6,500; monthly payments, $165; life insurance, $40,000.

WHAT YOU ARE TO DO

Complete the form with the information provided. Use the current date. If no information has been given, the blank is to be omitted. Sign the form.

State Bank   INSTALLMENT LOAN APPLICATION

I wish to apply for $______ which I can repay in______monthly payments.

This amount is to be used for______

My Full Name______ (PRINT) Wife's First Name______ Home Phone______

Home Address______ (NUMBER AND STREET) (CITY) (STATE) How Long______Years______Months

Previous Address______ (NUMBER AND STREET) (CITY) (STATE) How Long______Years______Months

Age______ Single ☐ Married ☐ Number of Dependents Including Wife______ I Prefer My Payments on the 1st ☐ 11th ☐ 21st ☐ 6th ☐ 16th ☐ 26th ☐

INCOME STATISTICS (Employment or Business)

Employed by______

Address______

Phone No.______ How Long There______Years______Months

Position______ Department______

Kind of Business______

Monthly Salary $______ Other Income $______

Source of Other Income______

Previous Employer______

Address______ How Long There______

Wife Employed by______

Phone No.______ How Long There______Years______Months

Position______ Monthly Salary $______

TWO CLOSE RELATIVES (Not Living With You)

Name______ (RELATIONSHIP)

Address______

Name______ (RELATIONSHIP)

Address______

BANK REFERENCES

Bank Account With______ (NAME OF BANK) ☐ Checking ☐ Savings

Title of Account______

Credit Union______ (NAME OF COMPANY) Value of Shares $______

RENT PAID OR REAL ESTATE OWNED

Location______ My Valuation $______

Title in Name of______

1st Mortgage $______ 2nd Mortgage $______

Mortgage Held by______

Mortgage Payment or—If You Rent—Rent per Month $______

Other Real Estate Owned______

Automobile Owned______ (YEAR) (MAKE) (MODEL)

Financed by______

Amount Owing $______ Monthly Payments $______

Stocks and Bonds Owned — Value $______

Description______

LIFE INSURANCE

Amount $______ Loans $______

LOANS, DEBTS AND ACCOUNTS OWED — OTHER THAN ABOVE (Include Any With Credit Unions)

	NAME OF BANK, COMPANY OR INDIVIDUAL	UNPAID BALANCE	MONTHLY PAYMENT
1.		$	$
2.		$	$
3.		$	$
4.		$	$
5.		$	$

Any other information or references that you care to submit______

The information above is true and complete and is given to induce you to grant credit to the undersigned.

Date______ Signed______

NAME: ______ JOB 16

I wish to apply for $___________ which I can repay in_____________monthly payments.

This amount is to be used for____

My Full Name_____________________________________   Wife's First Name__________   Home Phone__________
(PRINT)

Home Address_____________________________________   How Long _____Years_____Months
(NUMBER AND STREET)   (CITY)   (STATE)

Previous Address_________________________________   How Long _____Years_____Months
(NUMBER AND STREET)   (CITY)   (STATE)

Age__________   Single ☐   Married ☐   Number of Dependents Including Wife________________   I Prefer My Payments on the   1st ☐   11th ☐   21st ☐   6th ☐   16th ☐   26th ☐

INCOME STATISTICS (Employment or Business)

Employed by______________________________________

Address____

Phone No.__________   How Long There_____Years_____Months

Position________________________Department__________

Kind of Business_________________________________

Monthly Salary $__________   Other Income $__________

Source of Other Income___________________________

Previous Employer________________________________

Address__________   How Long There__________

Wife Employed by_________________________________

Phone No.__________   How Long There_____Years_____Months

Position__________   Monthly Salary $__________

TWO CLOSE RELATIVES (Not Living With You)

Name________________________________   (RELATIONSHIP)

Address____

Name________________________________   (RELATIONSHIP)

Address____

BANK REFERENCES

Bank Account With__________________________   ☐ Checking   ☐ Savings
(NAME OF BANK)

Title of Account_________________________________

Credit Union__________   Value of Shares $__________
(NAME OF COMPANY)

RENT PAID OR REAL ESTATE OWNED

Location__________   My Valuation $__________

Title in Name of_________________________________

1st Mortgage $__________   2nd Mortgage $__________

Mortgage Held by_________________________________

Mortgage Payment or—If You Rent—Rent per Month $__________

Other Real Estate Owned__________________________

Automobile Owned__________   (YEAR)   (MAKE)   (MODEL)

Financed by______________________________________

Amount Owing $__________   Monthly Payments $__________

Stocks and Bonds Owned — Value $__________

Description_______________________________________

LIFE INSURANCE

Amount $__________   Loans $__________

LOANS, DEBTS AND ACCOUNTS OWED — OTHER THAN ABOVE (Include Any With Credit Unions)

NAME OF BANK, COMPANY OR INDIVIDUAL	UNPAID BALANCE	MONTHLY PAYMENT
1.	$	$
2.	$	$
3.	$	$
4.	$	$
5.	$	$

Any other information or references that you care to submit__________________________

____

____

The information above is true and complete and is given to induce you to grant credit to the undersigned.

Date__________________________   Signed__________________________

JOB 17
Statement

- -

WHAT YOU SHOULD KNOW

Charge customer: Mr. J. D. Dungan, Route 2, Plano, Texas 75074-1452.

DAY OF MONTH	REFERENCES	CHARGES	CREDITS
2	#404 Cotton Seed	$246.00	
8	Cash		$200.00
16	#43 Fertilizer	137.50	
29	Spraying for disease:		
	Solution	537.50	
	Labor	320.00	

WHAT YOU ARE TO DO

Itemize the monthly charges to be sent to Mr. Dungan. Calculate the balance due and list it at the bottom of the statement. Use previous month of current year.

STATEMENT

DATE	REFERENCE	CHARGES	CREDITS

NAME: _______________________________   JOB 17

DATE	REFERENCE	CHARGES	CREDITS

NAME: _______________________ JOB 17

- -

WHAT YOU SHOULD KNOW

Wilson High School, 22000 Washington Boulevard, Toledo, Ohio 43624-6438, has ordered the following supplies:

REAMS	STYLE #	DESCRIPTION	PRICE
50	1123	16# Duplicator paper 8½ × 11 in. white	@ $5.06
20	1126	20# Mimeograph paper 8½ × 11 in. white	@ 5.16
10	1130	20# Mimeograph paper 8½ × 11 in. canary	@ 5.56
10	1131	20# Mimeograph paper 8½ × 11 in. blue	@ 5.56
10	1132	20# Mimeograph paper 8½ × 11 in. green	@ 5.56
10	1133	20# Mimeograph paper 8½ × 11 in. pink	@ 5.56
30	1152	20# Offset paper 8½ × 11 in. white	@ 5.06
10	1136	16# Mimeograph paper 8½ × 14 in. white	@ 5.26

The Customer Order No. is 7259, Invoice No. 1003. The salesman's number is 183, and the routing is Mohawk Truck. The quantity ordered and the quantity shipped are in agreement. Nothing was back-ordered.

WHAT YOU ARE TO DO

Figure the extensions for all items. Shipping charges amount to $15.50. The credit terms are 2/10, E.O.M. Net. The invoice will be paid within the discount period; therefore, compute the discount. E.O.M. Net means that if the invoice is not paid within the discount period, the full amount of the invoice is due on or before the end of the month.

INVOICE NUMBER

DATE

SOLD TO

SHIPPED TO

YOUR ORDER NUMBER

TERMS: 10TH E.O.M. NET THEREAFTER

SHIPPED VIA

SALESMAN

STYLE NUMBER	DESCRIPTION	QUANTITY	PRICE	AMOUNT

We hereby certify that these goods were produced in compliance with all applicable requirements of sections 6 and 12 of the Fair Labor Standards Act as amended and of regulations and orders of the United States Department of Labor issued under section 14 thereof.

Continuing guarantee under the Flammable Fabrics Act and the Fiber Products Identification Act and under Federal Trade Commission.

"PRICING POLICY" – LISTED PRICES OF ALL ITEMS ARE SUBJECT TO CHANGE TO REFLECT COSTS AT THE TIME OF DELIVERY. **TRANSPORTATION CHARGES NOT SUBJECT TO CASH DISCOUNT**

NAME: _______________________________ JOB 18

INVOICE NUMBER

DATE

SOLD TO

SHIPPED TO

YOUR ORDER NUMBER	TERMS: 10TH E.O.M. NET THEREAFTER	SHIPPED VIA	SALESMAN

STYLE NUMBER	DESCRIPTION	QUANTITY	PRICE	AMOUNT

We hereby certify that these goods were produced in compliance with all applicable requirements of sections 6, 7 and 12 of the Fair Labor Standards Act as amended and of regulations and orders of the United States Department of Labor issued under section 14 thereof.

Continuing guarantee under the Flammable Fabrics Act and the Fiber Products Identification Act and All Textile Fiber Commission.

"PRICING POLICY" – LISTED PRICES OF ALL ITEMS ARE SUBJECT TO CHANGE TO REFLECT COSTS AT THE TIME OF DELIVERY. **TRANSPORTATION CHARGES NOT SUBJECT TO CASH DISCOUNT**

NAME: _________________________________ **JOB 18** 107

WHAT YOU SHOULD KNOW

Royal High School, 4509 Smith Road, Dearborn, Michigan 48124-3663, has placed an order for equipment and supplies. The order was received yesterday and is being filled for shipment via Railway Express from Detroit tomorrow. The shipment includes:

	MODEL	SERIAL #	UNIT PRICE	AMOUNT
1 Microcomputer 48K, 1 Disk	3	26-1065	$1,395.00	$1,395.00
1 Microcomputer 48K 1 Disk	3	26-1066	1,395.00	1,395.00
1 Microcomputer 48K, 1 Disk	3	26-1067	1,395.00	1,395.00
1 Line Printer	7	26-1167	399.00	399.00
1 Cable for Printer		26-1401	39.00	39.00
3 Diskettes (10 per package)		26-0406	39.95	119.85
10 White Fanfold 9½ Paper		26-1423	7.95	79.50

The Customer's Order No. is 8205, dated one week ago. The Requisition No. is 5248, Invoice No. is 3972. The salesman is Harry Black.

WHAT YOU ARE TO DO

List each item separately. Use the current date for the invoice date. Type a line after the last item and show the total amount of the invoice.

CUSTOMER'S ORDER NO. & DATE	REQUISITION NO.	DATE ORDER REC'D	SHIPPED FROM	DATE SHIPPED	INVOICE DATE	
SALESMAN	ACCOUNT AT				VIA	

SOLD TO

SHIPPED TO

TERMS
NET 30 DAYS

QUANTITY		MODEL	SERIAL NO.	UNIT PRICE	AMOUNT

113

CUSTOMER'S ORDER NO. & DATE	REQUISITION NO.	DATE ORDER REC'D	SHIPPED FROM	DATE SHIPPED	INVOICE DATE	
SALESMAN	ACCOUNT AT				VIA	

SOLD
TO

SHIPPED TO

TERMS
NET 30 DAYS

QUANTITY		MODEL	SERIAL NO.	UNIT PRICE	AMOUNT

WHAT YOU SHOULD KNOW

This form is an application for credit for a charge account at a typical department store. The following information is needed:

Name, Jerry John Barrymore; address, 4122 Sycamore, Dallas, Texas 75204-8465; years lived at this address, 4; mortgage per month, $475; telephone number, (214) 282-2455; age, 27; marital status, married; dependents, 2; employer, Lone Star Gas Company, 4200 Jackson Street; telephone number of employer, (214) 272-4130; position, clerk; years employed, 6; superior's name, Mr. L. A. Johnson; income per month, $1,980; wife's first name, Doris; wife's employer, Texas Bank & Trust Company, 1200 Main Street; years employed, 3; position, secretary; employer's telephone number, (214) 867-3034; wife's income per month, $1,210; other sources of family income, none; previous account, none; name of bank, First National of Dallas (savings account and checking account); husband's nearest relative with whom not residing, Mr. & Mrs. J. J. Barrymore, 4900 W. North, Dallas; relationship, parents; all other creditors (all in Dallas area), Sanger-Harris (30-day account), balance due, $0, Kahn's (30-day account), balance due, $0, J. C. Penney's (30-day account), balance due, $0, Texas Bank & Trust Co., monthly payment, $475; balance due, $52,000.

WHAT YOU ARE TO DO

Complete the form. If no information has been given, the blank is to be omitted. Sign the form.

CREDIT APPLICATION

First Name	Middle	Last	Phone	Age	☐ Single ☐ Married	Dependents	Date	Limit

Address			Years at	Rent per mo. $	Mortgage per mo. $

City	Zone	State	Previous Address (If at present address less than 2 years)	Years at

Employer	Phone	Position or Occupation

Address	Years with	Badge No. or Superior's Name	Income per mo. $

Previous Employer (If at present job less than 2 years)	Address	Phone

Active Member of Armed Forces	Station	Branch	Rank	Service Number

Wife's first name	Employer	Address	Years with

Position	Employer's phone	Wife's income per mo. $	Other sources of family income	Amount per month $

Previous Acct. ☐ Yes ☐ No	Location	I bank at	☐ Savings acct. ☐ Checking acct.

Husband's nearest relative with whom not residing	Name	Address	Relationship

ALL OTHER CREDITORS: LIST ALL FIXED OBLIGATIONS, INSTALLMENT ACCOUNTS, FHA LOANS, ETC.

Name and Address	Month Paym't	Bal. Due	Name and Address	Month Paym't	Bal. Due

Interviewed by	Date	Signature	Date

CREDIT APPLICATION

First Name	Middle	Last	Phone	Age	☐ Single ☐ Married	Dependents	Date	Limit

Address			Years at	Rent per mo. $	Mortgage per mo. $	

City	Zone	State	Previous Address (If at present address less than 2 years)	Years at

Employer		Phone	Position or Occupation

Address		Years with	Badge No. or Superior's Name	Income per mo. $

Previous Employer (If at present job less than 2 years)	Address	Phone

Active Member of Armed Forces	Station	Branch	Rank	Service Number

Wife's first name	Employer	Address	Years with

Position	Employer's phone	Wife's income per mo. $	Other sources of family income	Amount per month $

Previous Acct. ☐ Yes ☐ No	Location	I bank at	☐ Savings acct. ☐ Checking acct.

Husband's nearest relative with whom not residing	Name	Address	Relationship

ALL OTHER CREDITORS: LIST ALL FIXED OBLIGATIONS, INSTALLMENT ACCOUNTS, FHA LOANS, ETC.

Name and Address	Month Paym't	Bal. Due	Name and Address	Month Paym't	Bal. Due

Interviewed by	Date	Signature	Date

NAME: _______________________________ JOB 20

Accounts Receivable Master File Worksheet

WHAT YOU SHOULD KNOW

Your Sales Department notifies the Data Processing Department of the following: (a.) additions of new customer accounts, (b.) changes of information in current accounts, and (c.) deletions of inactive accounts.

The information is typed on an Accounts Receivable Master File Worksheet. The special instructions below should be noted before typing:

- Characters of information should not exceed the number in parentheses. For example, the customer name should not exceed 22 characters including spaces; therefore, Johnston Computer Corporation would be typed as Johnston Computer Corp.

- Punctuation is omitted in the Name and Address lines.

- Address Line One: type the division name, if applicable, or Accounts Payable Department.

- Address Line Two: type the customer's street address.

- Address Line Three: type the name of the city.

- Nine-digit ZIP codes and telephone numbers are typed without hyphens.

- Dates are typed as six-digit numbers without slashes separating month, day, and year. Leading zeros are used for any month or day whose number does not take two digits. For example, the date July 8, 1983 would be typed as 070883.

- Account balances are typed using commas and decimal points. Dollar amounts are typed in this way:

$$\_\ \_\ \_\ \underline{8}\ \underline{7}\ \underline{3}\ \cdot\ \underline{5}\ \underline{7}\ (9)$$

Type the two Accounts Receivable Master File Worksheets using the information supplied by one of your salespeople. Remember to sign and date the form when you have completed it. Use the current date.

New Account:

Customer number: *4053918*   Salesperson's number *32*

Name: *Microcomputers for Fun*

Address: *362 N. Eagle Way*

Salt Lake, Utah 84108-0708

Telephone Number: *(801) 520-6515*

Credit Limit: *40,000*   Service Charge: *Y*

Price Code: *4*   Discount Code: *2*

Delete Account:

Customer number: *2678165*   Salesperson's number: *32*

Name: *Smythe and Jones Co.*

Address: *402 W. Freedom Drive*

Laramie, Wyoming 82070-1105

Account Balance: *0*   Date of last sale: *5/8/81*

□ NEW ACCOUNT

□ CHANGE INFORMATION

□ DELETE ACCOUNT

***** ACCOUNTS RECEIVABLE MASTER FILE WORKSHEET *****

CUSTOMER NUMBER   _ _ _ _ _ _ _ (7)

CUSTOMER NAME   _ (22)

ADDRESS LINE ONE   _ (22)

ADDRESS LINE TWO   _ (22)

ADDRESS LINE THREE   _ (22)

ZIP CODE   _ _ _ _ _ _ _ _ _ (9)

STATE CODE   _ _ (2)

TELEPHONE NUMBER   _ _ _ _ _ _ _ _ _ _ (10)

CURRENT MONTH BALANCE   _ _ _ _ _ _ _ _ _ (9)

CURRENT DUE BALANCE   _ _ _ _ _ _ _ _ _ (9)

30-DAY BALANCE   _ _ _ _ _ _ _ _ _ (9)

60-DAY BALANCE   _ _ _ _ _ _ _ _ _ (9)

90-DAY BALANCE   _ _ _ _ _ _ _ _ _ (9)

ACCOUNT BALANCE   _ _ _ _ _ _ _ _ _ (9)

CREDIT LIMIT   _ _ _ _ _ _ _ (7)

SALES YEAR-TO-DATE   _ _ _ _ _ _ _ _ _ (9)

RETURNS YEAR-TO-DATE   _ _ _ _ _ _ _ _ _ (9)

SALESMAN NUMBER   _ _ (2)

DATE LAST PAYMENT   _ _ _ _ _ _ (6)

DATE LAST SALE   _ _ _ _ _ _ (6)

BALANCE LAST STATEMENT _ _ _ _ _ _ _ _ _ (9)

SERVICE CHARGE   _ (1) Y OR N

PRICE CODE   _ (1)

DISCOUNT CODE   _ (1)

____
SIGNATURE OF TYPIST   DATE

☐ NEW ACCOUNT

☐ CHANGE INFORMATION

☐ DELETE ACCOUNT

***** ACCOUNTS RECEIVABLE MASTER FILE WORKSHEET *****

CUSTOMER NUMBER _ _ _ _ _ _ _ (7)

CUSTOMER NAME _ (22)

ADDRESS LINE ONE _ (22)

ADDRESS LINE TWO _ (22)

ADDRESS LINE THREE _ (22)

ZIP CODE _ _ _ _ _ _ _ _ _ (9)

STATE CODE _ _ (2)

TELEPHONE NUMBER _ _ _ _ _ _ _ _ _ _ (10)

CURRENT MONTH BALANCE _ _ _ _ _ _ _ _ _ (9)

CURRENT DUE BALANCE _ _ _ _ _ _ _ _ _ (9)

30-DAY BALANCE _ _ _ _ _ _ _ _ _ (9)

60-DAY BALANCE _ _ _ _ _ _ _ _ _ (9)

90-DAY BALANCE _ _ _ _ _ _ _ _ _ (9)

ACCOUNT BALANCE _ _ _ _ _ _ _ _ _ (9)

CREDIT LIMIT _ _ _ _ _ _ _ (7)

SALES YEAR-TO-DATE _ _ _ _ _ _ _ _ _ (9)

RETURNS YEAR-TO-DATE _ _ _ _ _ _ _ _ _ (9)

SALESMAN NUMBER _ _ (2)

DATE LAST PAYMENT _ _ _ _ _ _ (6)

DATE LAST SALE _ _ _ _ _ _ (6)

BALANCE LAST STATEMENT _ _ _ _ _ _ _ _ _ (9)

SERVICE CHARGE _ (1) Y OR N

PRICE CODE _ (1)

DISCOUNT CODE _ (1)

SIGNATURE OF TYPIST   DATE

Salesman's Report on Today's Extension Selling

WHAT YOU SHOULD KNOW

Mr. Dan Cook, District Manager of Toledo Branch, accompanied by Mr. James Hill, called on the purchasing agents of ten companies today and made his report as follows:

INDIVIDUAL OR FIRM	ADDRESS	TRIAL OBTAINED		FUTURE PROSPECT	
		Yes	No	Yes	No
Central Ohio Paper Company	126 North Ontario St. Toledo, OH 43624-5231	x		x	
Cousino Visual Edu. Service	1945 Franklin Avenue Toledo, OH 43624-5161		x		x
Adjusto Equipment Co.	515 Conneaut Bowling Green, OH 43402 7294	x		x	
Evans Office Equipment	723 Cory Street Findlay, OH 45840-4315	x		x	
Remington Rand Corp.	527 W. Woodruff Avenue Toledo, OH 43624-3910		x		x
Goodremont's Office	1846 E. Sylvania Ave. Toledo, OH 43612-1413	x		x	
Duplicating Machines	136 West Woodruff Toledo, OH 43624-0762		x		x
Monroe Calculating Co.	118 North Superior Toledo, OH 43604-1091	x		x	

(Continued on following page)

Wood County Press	134 East Wooster St. Bowling Green, OH 43402-1225	x	x
Miller-Bryant-Pierce	3141 Monroe Toledo, OH 43606-1455	x	x

WHAT YOU ARE TO DO

Complete the form. Each address will take two typed lines; you will have to use your variable line spacer to fit each address within the allotted space.

SALESMAN'S REPORT ON TODAY'S EXTENSION SELLING

☐ District Manager
☐ Market Research, E.O.
☐ Your file

Name: ______ {District: / Branch: ______ Paired With: ______ Date: ______

LIST HERE ANY SALES MADE THIS MONTH DIRECTLY OR INDIRECTLY RESULTING FROM THIS OR ANY PREVIOUS MONTH'S EXTENSION SELLING.

INDIVIDUAL OR FIRM	ADDRESS	TRIAL OBTAINED Yes	TRIAL OBTAINED No	FUTURE PROSPECT Yes	FUTURE PROSPECT No

Units Sold

Name of Account

SALESMAN'S REPORT ON TODAY'S EXTENSION SELLING

☐ District Manager
☐ Market Research, E.O.
☐ Your file

Name: District: Branch: Paired With: Date:

LIST HERE ANY SALES MADE **THIS MONTH** DIRECTLY OR INDIRECTLY RESULTING FROM THIS OR ANY PREVIOUS MONTH'S EXTENSION SELLING.

INDIVIDUAL OR FIRM	ADDRESS	TRIAL OBTAINED Yes	No	FUTURE PROSPECT Yes	No	# Units Sold / Name of Account

---

WHAT YOU SHOULD KNOW

Your Sales Department keeps a record of the companies your sales-people have visited during the month. The salespeople give you the information, which you type on the Sales Call Report form. These reports are valuable reminders to the salespeople.

WHAT YOU ARE TO DO

Type the two Sales Call Reports using the information supplied by your salesperson, Jim Craig.

Date of call (use Friday's date of the previous week); Mifflin Office Equipment; 921 Cord Street; Indianapolis, Indiana 46224-4930; Phone: 381-9120; active customer; Ms. Carla Jackson, Purchasing Agent; send new catalog and price list; follow-up date (3 weeks from last visit).

Date of call (same as above); prospect; Automated Office Supplies; 327 Kildare Avenue; Indianapolis, Indiana 46218-8278; Phone: 384-9090; Mr. Earl Miller, Owner; Remarks: Very interested in word processing equipment. Call next Friday to arrange for luncheon meeting with the sales manager. Add to mailing list.

SALES CALL REPORT

SALESPERSON _____________________________________ DATE OF CALL ___ / ___ / ___

☐ ACTIVE CUSTOMER ☐ INACTIVE CUSTOMER ☐ NEW CUSTOMER ☐ PROSPECT

COMPANY _________________________________ PHONE _____________

ADDRESS ____

CITY _________________________ STATE ____________ ZIP __________

TALKED TO _________________________________ TITLE _____________

SOLD TO _______________________________ AMOUNT OF ORDER _______

REMARKS AND SPECIAL NOTES

____

____

____

____

FOLLOW-UP DATE ___ / ___ / ___

ADD TO MAILING LIST: ☐ YES ☐ NO   SEND: ☐ CATALOG ☐ PRICE LIST ☐ OTHER

SALES CALL REPORT

SALESPERSON _______________________________ DATE OF CALL ___/___/___

☐ ACTIVE CUSTOMER ☐ INACTIVE CUSTOMER ☐ NEW CUSTOMER ☐ PROSPECT

COMPANY _______________________________ PHONE _______________

ADDRESS ____

CITY _______________________ STATE _____________ ZIP __________

TALKED TO _______________________ TITLE _______________

SOLD TO _______________________ AMOUNT OF ORDER __________

REMARKS AND SPECIAL NOTES

____

____

____

____

FOLLOW-UP DATE ___/___/___

ADD TO MAILING LIST: ☐ YES ☐ NO   SEND: ☐ CATALOG ☐ PRICE LIST ☐ OTHER

NAME: _______________________ JOB 23   137

JOB 24
Requisition for Material

- -

WHAT YOU SHOULD KNOW

The Research Department in Denver, Colorado, has requisitioned from the Purchasing Department in Findlay, Ohio, the following items, which are needed four weeks from today. The vendor is the A. B. Dick Company.

4 quire	Newspaper Stencils, 960L-1071
6 lbs.	#3200 Black Paste Ink
2 lbs.	#3204 Brown Paste Ink
2 pts.	Offset Hand Cleaner, #4-4921
4 pkgs.	Cotton Pads
3 pkgs.	Offset Clean-Up Mats
100	#5000 Offset Masters
1 pkg.	Aluminum Plates, #84-4967
3 qts.	Photoplax Developer, #84-4969
2 qts.	Photocopy Developer, #74-2000

Additional Information: ship via Railroad Freight in care of Fred Stone; 10018 South Fourth Street; ZIP Code 80201-4305; requisition number 2905; Sheet #4; Field Request #359; and commodity Group #16.

WHAT YOU ARE TO DO

Complete the requisition except where signatures are needed. Use the current date.

REQUISITION FOR MATERIAL

To: Purchasing Agent

Ship to ____________________________ , Dept.

Town ___________ County ___________ State ___________

Care Of: ____________________________

Street Address ____________________________

Ship Via: Prepaid ___ Truck Freight ___ Railway Express ___ Will Call ___
Railroad Freight ___ Parcel Post ___ Vendor Delivery ___

Name Of Carrier At Destination ____________________________

Requisition No. ____________________________

Sheet No. ____________________________

Date ___________ , 19 ___

INSTRUCTIONS

Use This Form When Ordering Material Through Findlay Purchasing Division.

Copies: White—Mail to Findlay Office.
Pink—Retain in Division.
Canary—Retain in District.
Green—Retained by Originator.

Descriptions must be Complete, Accurate and in Detail.

Commodity Group No. ___ Date Material Needed ___________ Field Request No. ___________

QUANTITY	DESCRIPTION OF MATERIAL	VENDOR	TO BE USED FOR—REMARKS

PURCHASING DIVISION USE ONLY

PURCHASING DIVISION USE ONLY

Signed ____________________ Signed ____________________ Approved ____________________

REQUISITION FOR MATERIAL

To: Purchasing Agent

Ship to ___________________________ , Dept.

Town ________ County ________ State ________

Care Of: ___________________________

Street Address ___________________________

Ship Via: Prepaid ____ Truck Freight ____ Railroad Freight ____ Failway Express ____ Parcel Post ____ Will Call ____ Vendor Delivery ____

Name Of Carrier At Destination ___________________________

Requisition No. ___________________________

Sheet No. ___________________________

Date ____________ , 19

INSTRUCTIONS

Use This Form When Ordering Material Through Findlay Purchasing Division.

Copies: White—Mail to Findlay Office.
Pink—Retain in Division.
Canary—Retain in District.
Green—Retained by Originator.

Descriptions must be Complete, Accurate and in Detail.

Commodity Group No. ____ Date Material Needed ____________ Field Request No. ____________

QUANTITY	DESCRIPTION OF MATERIAL	VENDOR	TO BE USED FOR—REMARKS

PURCHASING DIVISION USE ONLY

Signed ____________ Signed ____________ Approved ____________

Requisition for Supplies

WHAT YOU SHOULD KNOW

Many businesses require department managers to submit signed requisitions to the Purchasing Department before supplies are ordered. This helps the business to maintain sufficient quantities of supplies in inventory without overstocking.

WHAT YOU ARE TO DO

You are the secretary to Mr. Leo Haley, Manager of the Accounting Department. He has asked you to order the following items:

12 pkgs	251010L Correctable Film Ribbons	4.98
30	1010L Lift-Off Tape	.63
10 boxes	152L Hanging File Folders-legal	7.09
10	157923L Desktop Legal Tray	7.89

Using today's date, type the Requisition; have the order shipped to your attention and charged to Supplies Expense, Account #6532. Mr. Haley has authorized you to approve and sign the requisition. Order from the Swanson Supplies Company, 2201 Spruce Street, Montgomery, Alabama 36106-6025.

REQUISITION FOR SUPPLIES

Date _________________________ Department _________________________________

TO ____

____

____

____

Item	Quantity	Description	Price
1			
2			
3			
4			

Ship to the attention of: _________________________ Charge to account #: _________________

Approved by: _________________________

REQUISITION FOR SUPPLIES

Date _______________________ Department _______________________________

TO ____

____

____

____

Item	Quantity	Description	Price
1			
2			
3			
4			

Ship to the attention of: _______________________   Charge to account #: _______________

Approved by: _______________________

Authorized Payroll Deductions

WHAT YOU SHOULD KNOW

On being hired for her first job, Joan Carol White, 625 Loop Road, Juneau, Alaska 99501-1924, found it was necessary to give the Payroll Department certain information concerning deductions from her paycheck. These deductions included group insurance, withholding tax, company recreation club, and U.S. Savings Bonds. Ms. White was employed in the Accounting Department one week ago today. Her date of birth is September 8, 1965. Beneficiary is John B. White, her father. His address is 72 Commerce, Augusta, Maine 04330-2719. She is single, with no dependents. She wishes to participate in the Basic Plan of Group Insurance. She wishes to participate in the Payroll Deduction Plan for U.S. Savings Bonds, with her father as beneficiary, at the deduction rate of $15.00 per pay period for a $100 bond. Her account number will be 7564.

WHAT YOU ARE TO DO

Using the current date, fill in all the blanks that apply to this employee.

AUTHORIZED PAYROLL DEDUCTIONS

DATE

EMPLOYEE

DEPT.

DATE OF BIRTH

DATE EMPLOYED

GROUP CENSUS, ENROLLMENT AND RECORD CARD

NAME OF BENEFICIARY—THE RIGHT TO CHANGE THE BENEFICIARY IS RESERVED

RELATIONSHIP TO EMPLOYEE

ADDRESS OF BENEFICIARY

☐ SINGLE EMPLOYEE

☐ EMPLOYEE WITH DEPENDENT(S) AS DEFINED BELOW

THE ABOVE DESIGNATION OF BENEFICIARY REVOKES ANY PRIOR DESIGNATION INCONSISTENT THEREWITH.
THE TERM "DEPENDENT" IS LIMITED TO (A) THE EMPLOYEE'S UNMARRIED CHILDREN UNDER NINETEEN (19) YEARS OF AGE AND (B) THE EMPLOYEE'S SPOUSE, RESIDING IN THE UNITED STATES OR CANADA.
NOTE: AN EMPLOYEE IS NOT ENTITLED TO BENEFITS FOR A SPOUSE WHO IS INSURED AS AN EMPLOYEE NOR FOR A CHILD WHO IS AN EMPLOYEE OF THE EMPLOYER.

☐ BASIC PLAN

☐ OPTIONAL PLAN

INS. CLASS

NON - CONTRIBUTORY

SUBJECT TO MY RIGHT TO REVOKE THIS ORDER AT ANY TIME BY WRITTEN NOTICE. I HEREBY AUTHORIZE MY EMPLOYER TO COLLECT BY PAYROLL DEDUCTION OR OTHERWISE THE AMOUNT AS MAY FROM TIME TO TIME BE DETERMINED AND REQUIRED TO APPLY TOWARD THE PREMIUMS FOR THE FORMS OF GROUP INSURANCE PROVIDED FOR IN THE POLICY OR POLICIES OF GROUP INSURANCE ISSUED TO COMPANY BY THE TRAVELERS INSURANCE

SIGN HERE →

SIGN HERE →

EMPLOYEE'S WITHHOLDING EXEMPTION CERTIFICATE— HOW TO CLAIM YOUR WITHHOLDING EXEMPTIONS

EMPLOYEE:

File this form with your employer. Otherwise, he must withhold U. S. income tax from your wages without exemption.

EMPLOYER:

Keep this certificate with your records. If the employee is believed to have claimed too many exemptions, the District Director should be so advised.

1. If SINGLE, and you claim an exemption, write the figure "1"..................
   (a) If you claim BOTH of these exemptions, write
2. If MARRIED, one exemption   the figure "2"
   each for husband and wife(b) If you claim ONE of these exemptions, write
   if not claimed on another   the figure "1"
   certificate.   (c) If you claim NEITHER of these exemptions,
                                write "0"
3. Exemptions for age and blindness (applicable only to you and your wife but NOT to dependents:)
   (a) If you or your wife will be 65 years of age or older at the end of the year, and you claim this exemption, write the figure "1"; if BOTH will be 65 or older, and you claim both of these exemptions, write the figure "2"...............
   (b) If you or your wife are blind, and you claim this exemption, write the figure "1"; if BOTH are blind, and you claim both of these exemptions, write the figure "2"...............
4. If you claim exemptions for one or more dependents, write the number of such exemptions. (Do not claim exemption for a dependent unless you are qualified under instruction 3 on other side.)...............
5. Add the number of exemptions which you have claimed above and write the total. ....

I CERTIFY that the number of withholding exemptions claimed on this certificate does not exceed the number to which I am entitled.

SIGN HERE →

I WANT TO BE A MEMBER OF THE A. R. C. THIS IS YOUR AUTHORITY TO DEDUCT DUES OF $1.00 SEMI-ANNUALLY FROM MY PAY CHECK UNTIL OTHERWISE NOTIFIED.

SIGN HERE →

PAYROLL DEDUCTION PLAN FOR U.S. SAVINGS BONDS

ACCOUNT NUMBER

DEDUCT EACH PAY PERIOD $

1.25 ☐ 12.50 ☐ 25.00 ☐
3.75 ☐ 18.00 ☐ 37.50 ☐
6.25 ☐ 18.75 ☐ 75.00 ☐

OTHER

MATURITY VALUE ($)   25 ☐ 50 ☐ 100 ☐   OTHER

MR. / MRS. / MISS   OWNER NAME
FIRST   MIDDLE INITIAL   LAST

NUMBER   STREET

CITY   STATE

MR. ☐ CO-OWNER OR ☐ BENEFICIARY NAME
MRS. / MISS   FIRST   MIDDLE INITIAL   LAST

NUMBER   STREET

CITY   STATE

AUTHORIZED PAYROLL DEDUCTIONS

DATE

EMPLOYEE

DEPT.

DATE OF BIRTH

DATE EMPLOYED

GROUP CENSUS, ENROLLMENT AND RECORD CARD

NAME OF BENEFICIARY—THE RIGHT TO CHANGE THE BENEFICIARY IS RESERVED | RELATIONSHIP TO EMPLOYEE

ADDRESS OF BENEFICIARY

☐ SINGLE EMPLOYEE   ☐ EMPLOYEE WITH DEPENDENT(S) AS DEFINED BELOW

THE ABOVE DESIGNATION OF BENEFICIARY REVOKES ANY PRIOR DESIGNATION INCONSISTENT THEREWITH.
THE TERM "DEPENDENT" IS LIMITED TO (A) THE EMPLOYEE'S UNMARRIED CHILDREN UNDER NINETEEN (19) YEARS OF AGE AND (B) THE EMPLOYEE'S SPOUSE, RESIDING IN THE UNITED STATES OR CANADA.
NOTE: AN EMPLOYEE IS NOT ENTITLED TO BENEFITS FOR A SPOUSE WHO IS INSURED AS AN EMPLOYEE NOR FOR A CHILD WHO IS AN EMPLOYEE OF THE EMPLOYER.

☐ BASIC PLAN

☐ OPTIONAL PLAN | INS. CLASS

NON - CONTRIBUTORY

SUBJECT TO MY RIGHT TO REVOKE THIS ORDER AT ANY TIME BY WRITTEN NOTICE, I HEREBY AUTHORIZE MY EMPLOYER TO COLLECT BY PAYROLL DEDUCTION OR OTHERWISE THE AMOUNT AS MAY FROM TIME TO TIME BE DETERMINED AND REQUIRED TO APPLY TOWARD THE PREMIUMS FOR THE FORMS OF GROUP INSURANCE PROVIDED FOR IN THE POLICY OR POLICIES OF GROUP INSURANCE ISSUED TO COMPANY BY THE TRAVELERS INSURANCE

SIGN HERE ▶

SIGN HERE ▶

EMPLOYEE'S WITHHOLDING EXEMPTION CERTIFICATE— HOW TO CLAIM YOUR WITHHOLDING EXEMPTIONS

EMPLOYEE:
File this form with your employer. Otherwise, he must withhold U. S. income tax from your wages without exemption.

EMPLOYER:
Keep this certificate with your records. If the employee is believed to have claimed too many exemptions, the District Director should be so advised.

1. If SINGLE, and you claim an exemption, write the figure "1" ....
2. If MARRIED, one exemption the figure "2" each for husband and wife (b) If you claim ONE of these exemptions, write if not claimed on another the figure "1" certificate.
    (a) If you claim BOTH of these exemptions, write.
    (b) If you claim ONE of these exemptions, write
    (c) If you claim NEITHER of these exemptions, write "0"
3. Exemptions for age and blindness (applicable only to you and your wife but NOT to dependents:)
    (a) If you or your wife will be 65 years of age or older at the end of the year, and you claim this exemption, write the figure "1"; if BOTH will be 65 or older, and you claim both of these exemptions, write the figure "2" ....
    (b) If you or your wife are blind, and you claim this exemption, write the figure "1"; if BOTH are blind, and you claim both of these exemptions, write the figure "2" ....
4. If you claim exemptions for one or more dependents, write the number of such exemptions. (Do not claim exemption for a dependent unless you are qualified under Instruction 3 on other side.) ....
5. Add the number of exemptions which you have claimed above and write the total. ....

I CERTIFY that the number of withholding exemptions claimed on this certificate does not exceed the number to which I am entitled.

SIGN HERE ▶

I WANT TO BE A MEMBER OF THE A. R. C. THIS IS YOUR AUTHORITY TO DEDUCT DUES OF $1.00 SEMI-ANNUALLY FROM MY PAY CHECK UNTIL OTHERWISE NOTIFIED.

SIGN HERE ▶

PAYROLL DEDUCTION PLAN FOR U.S. SAVINGS BONDS

ACCOUNT NUMBER

DEDUCT EACH PAY PERIOD $

☐ 1.25 ☐ 12.50 ☐ 25.00
☐ 3.75 ☐ 18.00 ☐ 37.50
☐ 6.25 ☐ 18.75 ☐ 75.00

OTHER

MATURITY VALUE ($) ☐ 25 ☐ 50 ☐ 100 OTHER

MR. MRS. MISS | OWNER NAME | FIRST MIDDLE INITIAL LAST

NUMBER STREET

CITY STATE

MR. MRS. MISS | ☐ CO-OWNER OR ☐ BENEFICIARY NAME | FIRST MIDDLE INITIAL LAST

NUMBER STREET

CITY STATE

WHAT YOU SHOULD KNOW

This job will improve your ability to type numbers. Notice that the time is given in tenths of hours in addition to regular A.M. and P.M. times. This makes it easier to assign the time the machine is used to the job number. The machine number is 7070. Repeat the job number on each line of times indicated.

04.0–05.7	136958
05.8–05.9	65421
06.0–07.9	235012
08.0–09.0	268
09.1–10.4	13420
10.5–11.4	6892
11.5–12.8	13
12.9–15.2	56821
15.3–16.0	64284
16.0–17.0	6891
17.1–19.5	14
19.6–20.4	26
20.5–20.6	94382
20.7–24.0	146921

WHAT YOU ARE TO DO

Complete the form using the current date. Be sure that you type the correct number on each line. Proofread carefully! Use last Thursday's date as the Working Day.

MACHINE ANALYSIS

Date _______________________

Machine Number _____________________________   Working Day _____________________________

TIME		JOB NUMBER	TIME		JOB NUMBER	TIME		JOB NUMBER
4:00	04.00		8:00	08.00		12:N	12.00	
4:06	04.10		8:06	08.10		12:06	12.10	
4:12	04.20		8:12	08.20		12:12	12.20	
4:18	04.30		8:18	08.30		12:18	12.30	
4:24	04.40		8:24	08.40		12:24	12.40	
4:30	04.50		8:30	08.50		12:30	12.50	
4:36	04.60		8:36	08.60		12:36	12.60	
4:42	04.70		8:42	08.70		12:42	12.70	
4:48	04.80		8:48	08.80		12:48	12.80	
4:54	04.90		8:54	08.90		12:54	12.90	
5:00	05.00		9:00	09.00		1:00	13.00	
5:06	05.10		9:06	09.10		1:06	13.10	
5:12	05.20		9:12	09.20		1:12	13.20	
5:18	05.30		9:18	09.30		1:18	13.30	
5:24	05.40		9:24	09.40		1:24	13.40	
5:30	05.50		9:30	09.50		1:30	13.50	
5:36	05.60		9:36	09.60		1:36	13.60	
5:42	05.70		9:42	09.70		1:42	13.70	
5:48	05.80		9:48	09.80		1:48	13.80	
5:54	05.90		9:54	09.90		1:54	13.90	
6:00	06.00		10:00	10.00		2:00	14.00	
6:06	06.10		10:06	10.10		2:06	14.10	
6:12	06.20		10:12	10.20		2:12	14.20	
6:18	06.30		10:18	10.30		2:18	14.30	
6:24	06.40		10:24	10.40		2:24	14.40	
6:30	06.50		10:30	10.50		2:30	14.50	
6:36	06.60		10:36	10.60		2:36	14.60	
6:42	06.70		10:42	10.70		2:42	14.70	
6:48	06.80		10:48	10.80		2:48	14.80	
6:54	06.90		10:54	10.90		2:54	14.90	
7:00	07.00		11:00	11.00		3:00	15.00	
7:06	07.10		11:06	11.10		3:06	15.10	
7:12	07.20		11:12	11.20		3:12	15.20	
7:18	07.30		11:18	11.30		3:18	15.30	
7:24	07.40		11:24	11.40		3:24	15.40	
7:30	07.50		11:30	11.50		3:30	15.50	
7:36	07.60		11:36	11.60		3:36	15.60	
7:42	07.70		11:42	11.70		3:42	15.70	
7:48	07.80		11:48	11.80		3:48	15.80	
7:54	07.90		11:54	11.90		3:54	15.90	
8:00	08.00		12:N	12.00		4:00	16.00	

over

TIME		JOB NUMBER	TIME		JOB NUMBER
4:00	16.00		8:00	20.00	
4:06	16.10		8:06	20.10	
4:12	16.20		8:12	20.20	
4:18	16.30		8:18	20.30	
4:24	16.40		8:24	20.40	
4:30	16.50		8:30	20.50	
4:36	16.60		8:36	20.60	
4:42	16.70		8:42	20.70	
4:48	16.80		8:48	20.80	
4:54	16.90		8:54	20.90	
5:00	17.00		9:00	21.00	
5:06	17.10		9:06	21.10	
5:12	17.20		9:12	21.20	
5:18	17.30		9:18	21.30	
5:24	17.40		9:24	21.40	
5:30	17.50		9:30	21.50	
5:36	17.60		9:36	21.60	
5:42	17.70		9:42	21.70	
5:48	17.80		9:48	21.80	
5:54	17.90		9:54	21.90	
6:00	18.00		10:00	22.00	
6:06	18.10		10:06	22.10	
6:12	18.20		10:12	22.20	
6:18	18.30		10:18	22.30	
6:24	18.40		10:24	22.40	
6:30	18.50		10:30	22.50	
6:36	18.60		10:36	22.60	
6:42	18.70		10:42	22.70	
6:48	18.80		10:48	22.80	
6:54	18.90		10:54	22.90	
7:00	19.00		11:00	23.00	
7:06	19.10		11:06	23.10	
7:12	19.20		11:12	23.20	
7:18	19.30		11:18	23.30	
7:24	19.40		11:24	23.40	
7:30	19.50		11:30	23.50	
7:36	19.60		11:36	23.60	
7:42	19.70		11:42	23.70	
7:48	19.80		11:48	23.80	
7:54	19.90		11:54	23.90	
8:00	20.00		12:**M**	24.00	

MACHINE ANALYSIS

Date ______________________

Machine Number ______________________________   Working Day ______________________

TIME		JOB NUMBER	TIME		JOB NUMBER	TIME		JOB NUMBER
4:00	04.00		8:00	08.00		12:N	12.00	
4:06	04.10		8:06	08.10		12:06	12.10	
4:12	04.20		8:12	08.20		12:12	12.20	
4:18	04.30		8:18	08.30		12:18	12.30	
4:24	04.40		8:24	08.40		12:24	12.40	
4:30	04.50		8:30	08.50		12:30	12.50	
4:36	04.60		8:36	08.60		12:36	12.60	
4:42	04.70		8:42	08.70		12:42	12.70	
4:48	04.80		8:48	08.80		12:48	12.80	
4:54	04.90		8:54	08.90		12:54	12.90	
5:00	05.00		9:00	09.00		1:00	13.00	
5:06	05.10		9:06	09.10		1:06	13.10	
5:12	05.20		9:12	09.20		1:12	13.20	
5:18	05.30		9:18	09.30		1:18	13.30	
5:24	05.40		9:24	09.40		1:24	13.40	
5:30	05.50		9:30	09.50		1:30	13.50	
5:36	05.60		9:36	09.60		1:36	13.60	
5:42	05.70		9:42	09.70		1:42	13.70	
5:48	05.80		9:48	09.80		1:48	13.80	
5:54	05.90		9:54	09.90		1:54	13.90	
6:00	06.00		10:00	10.00		2:00	14.00	
6:06	06.10		10:06	10.10		2:06	14.10	
6:12	06.20		10:12	10.20		2:12	14.20	
6:18	06.30		10:18	10.30		2:18	14.30	
6:24	06.40		10:24	10.40		2:24	14.40	
6:30	06.50		10:30	10.50		2:30	14.50	
6:36	06.60		10:36	10.60		2:36	14.60	
6:42	06.70		10:42	10.70		2:42	14.70	
6:48	06.80		10:48	10.80		2:48	14.80	
6:54	06.90		10:54	10.90		2:54	14.90	
7:00	07.00		11:00	11.00		3:00	15.00	
7:06	07.10		11:06	11.10		3:06	15.10	
7:12	07.20		11:12	11.20		3:12	15.20	
7:18	07.30		11:18	11.30		3:18	15.30	
7:24	07.40		11:24	11.40		3:24	15.40	
7:30	07.50		11:30	11.50		3:30	15.50	
7:36	07.60		11:36	11.60		3:36	15.60	
7:42	07.70		11:42	11.70		3:42	15.70	
7:48	07.80		11:48	11.80		3:48	15.80	
7:54	07.90		11:54	11.90		3:54	15.90	
8:00	08.00		12:N	12.00		4:00	16.00	

over

PAGE 2

TIME		JOB NUMBER	TIME		JOB NUMBER
4:00	16.00		8:00	20.00	
4:06	16.10		8:06	20.10	
4:12	16.20		8:12	20.20	
4:18	16.30		8:18	20.30	
4:24	16.40		8:24	20.40	
4:30	16.50		8:30	20.50	
4:36	16.60		8:36	20.60	
4:42	16.70		8:42	20.70	
4:48	16.80		8:48	20.80	
4:54	16.90		8:54	20.90	
5:00	17.00		9:00	21.00	
5:06	17.10		9:06	21.10	
5:12	17.20		9:12	21.20	
5:18	17.30		9:18	21.30	
5:24	17.40		9:24	21.40	
5:30	17.50		9:30	21.50	
5:36	17.60		9:36	21.60	
5:42	17.70		9:42	21.70	
5:48	17.80		9:48	21.80	
5:54	17.90		9:54	21.90	
6:00	18.00		10:00	22.00	
6:06	18.10		10:06	22.10	
6:12	18.20		10:12	22.20	
6:18	18.30		10:18	22.30	
6:24	18.40		10:24	22.40	
6:30	18.50		10:30	22.50	
6:36	18.60		10:36	22.60	
6:42	18.70		10:42	22.70	
6:48	18.80		10:48	22.80	
6:54	18.90		10:54	22.90	
7:00	19.00		11:00	23.00	
7:06	19.10		11:06	23.10	
7:12	19.20		11:12	23.20	
7:18	19.30		11:18	23.30	
7:24	19.40		11:24	23.40	
7:30	19.50		11:30	23.50	
7:36	19.60		11:36	23.60	
7:42	19.70		11:42	23.70	
7:48	19.80		11:48	23.80	
7:54	19.90		11:54	23.90	
8:00	20.00		12:M	24.00	

Uncollected C.O.D. Sheet

- -

WHAT YOU SHOULD KNOW

Seven yearbooks have not been called for at the main store of Yearbook, Inc. It is necessary that the auditor be informed about those who have not met their commitments. The customers and their locations follow.

CUSTOMER	LOCATION	AMOUNT
Tom Cook	52 Featherstone Drive Little Rock, AR 72210-4923	$18.00
Lorie Hall	6 Tyler Street Little Rock, AR 72204-2262	18.00
Virginia Abbott	602 Greenfield Drive Little Rock, AR 72209-2436	18.00
Dick Bennett	1135 Yorkshire Place Little Rock, AR 72205-5110	18.00
Bob Jackson	35265 Dalewood Road Little Rock, AR 72207-7240	18.00
Shannon Laws	3502 Hopson Drive Little Rock, AR 72209-1964	18.00
Larry Garrett	1003 White Road Little Rock, AR 72211-3247	18.00

WHAT YOU ARE TO DO

Type names in alphabetic order. Use the first day of the current month as the date of charge for each customer. The amount is $18.00 in each case. Use two vertical spaces for the street address and the city and state. Double space between addresses.

UN-COLLECTED C.O.D. SHEET

Store ....

     C. O. D. Charge Slips are to be retained at Store Offices a reasonable time to allow for collection.

     On the last day of each month this sheet must be returned to Auditor listing all uncollected items.

     If you have no such items return sheet marked "NONE".

Customer	Location	Date of Charge	Amount

NAME: _________________________________ JOB 28

UN-COLLECTED C.O.D. SHEET

Store ....

       C. O. D. Charge Slips are to be retained at Store Offices a reasonable time to allow for collection.

       On the last day of each month this sheet must be returned to Auditor listing all uncollected items.

       If you have no such items return sheet marked "NONE".

Customer	Location	Date of Charge	Amount

NAME: _______________________________ JOB 28

Merchandise Received

WHAT YOU SHOULD KNOW

Yesterday a local store received a prepaid order, via parcel post, from the Ajax Equipment Company. The order weighed 18 pounds and consisted of:

1,000	#10 envelopes, $4\frac{1}{8} \times 9\frac{1}{2}$ in.
500	#6¾ envelopes, $3\frac{5}{8} \times 6\frac{1}{2}$ in.
1,000	3 × 5 in. file cards
500	letterheads, $8\frac{1}{2} \times 11$ in.
500	16# sulphite second sheets, $8\frac{1}{2} \times 11$ in.

WHAT YOU ARE TO DO

The Purchase Order number is 5437. The order arrived in three boxes and was delivered to the Supplies Department. Use the current date and fill in the appropriate blanks with the data given.

M E R C H A N D I S E   R E C E I V E D

From _______________________________   Date _______________   Purchase Order _______________

Via: Parcel Post ☐   Railway Express ☐   Our Pickup ☐   Shippers Deliveries ☐   Railroad _______________   Truck Line _______________

Date _______________   Pro. Nos. _______________   Weight _______________   No. of Pieces _______________   Prepaid ☐   Collect $ _______________

QUANTITIES	FORM NO.	TITLE / DESCRIPTION	BIN	SHELF-RACK

Delivered To _______________________________   Signed _______________________________

MERCHANDISE RECEIVED

From ___________________________   **Date** ___________________   **Purchase Order** ___________

Via: Parcel Post ☐   Railway Express ☐   Our Pickup ☐   Shippers Deliveries ☐   Railroad ___________________   Truck Line ___________________

Date ___________   Pro. Nos. ___________   Weight ___________   No. of Pieces ___________   Prepaid ☐   Collect $ ___________

QUANTITIES	FORM NO.	TITLE / DESCRIPTION	BIN	SHELF-RACK

Delivered To ___________________________________   **Signed** ___________________________________

NAME: ___________________________   JOB 29   173

JOB 30
Daily Absentee Report

- -

WHAT YOU SHOULD KNOW

The following information was delivered to the Payroll Department; Section 240; Unit: Production; Room 306; Date: current date.

Five employees, listed below, are absent today for the reasons given:

Judy Wright	Influenza
Susan Smith	Broken leg
Tom Young	Dental appointment
Sally Mayer	Sore throat
William Jones	Grandmother died

WHAT YOU ARE TO DO

Fill in blank spaces on the form, using the information given. Arrange the names in alphabetical order. Type an "X" in the appropriate space for each employee listed. Sign your name as first supervisor and omit the bottom line.

DAILY ABSENTEE REPORT

DELIVER TO: ______________________________   ROOM ______________

SECTION ______________________   UNIT ______________________   DATE ______________

NAME OF EMPLOYEE	LATE	HOSPITALIZED ILLNESS	NON-HOSP. ILLNESS	ILLNESS IN FAMILY	DEATH IN FAMILY	FUNERAL	EMPLOYEE MEDICAL APPOINTMENT	VACATION	OTHER

REMARKS

UNDER THIS HEADING ALWAYS SHOW:

1. NATURE OF ILLNESS.
2. PERTINENT FACTS REGARDING LATENESS, ILLNESS IN FAMILY, DEATH IN FAMILY, AND EMPLOYEE MEDICAL APPOINTMENT.
3. REASON FOR AN ABSENCE INDICATED AS "OTHER".

PLEASE COMPLETE AND PLACE IN OFFICE MAIL BEFORE 8:30 A. M.

____
SUPERVISOR

____
SUPERVISOR

NAME: ______________________________   JOB 30   177

DAILY ABSENTEE REPORT

DELIVER TO: ______________________________   ROOM ______________

SECTION ____________________   UNIT ______________   DATE ______________

NAME OF EMPLOYEE	LATE	HOSPITALIZED ILLNESS	NON-HOSP. ILLNESS	ILLNESS IN FAMILY	DEATH IN FAMILY	FUNERAL	EMPLOYEE MEDICAL APPOINTMENT	VACATION	OTHER	REMARKS

REMARKS

UNDER THIS HEADING ALWAYS SHOW:

1. NATURE OF ILLNESS.
2. PERTINENT FACTS REGARDING LATENESS, ILLNESS IN FAMILY, DEATH IN FAMILY, AND EMPLOYEE MEDICAL APPOINTMENT.
3. REASON FOR AN ABSENCE INDICATED AS "OTHER"

PLEASE COMPLETE AND PLACE IN OFFICE MAIL BEFORE 8:30 A. M.

____
SUPERVISOR

____
SUPERVISOR

JOB 31
Maintenance or Repair
Request

WHAT YOU SHOULD KNOW

John Robot, Room 205, finds that his electronic typewriter with a lift-off correction function fails to make complete corrections. He requests that it be repaired before next Monday and assigns the responsibility to the Machines Repair Section.

WHAT YOU ARE TO DO

Fill in all the blanks except the actual completion date. Use the current date as the date of request.

MAINTENANCE OR REPAIR REQUEST

Date_____________________ Room Number_____________

Requested by____

Description

____

____

____

____

____

____

____

____

To be Completed by_________________________________
Date

Assigned to____

Actual Completion Date_____________________________

MAINTENANCE OR REPAIR REQUEST

Date_________________________ Room Number___________

Requested by____

<u>Description</u>

____

____

____

____

____

____

____

____

To be Completed by___________________________________
Date

Assigned to____

Actual Completion Date_______________________________

Employee's Daily Time Card

WHAT YOU SHOULD KNOW

You are employee number 1368, working as a painter on the 2nd shift. You performed the following work during the shift:

START	WORK ORDER NUMBER	DEPT.	ACCT.	ACCT. NO.	DESCRIPTION
4:00	165	52	81	261	Paint hinges
5:25	1386	21	10	322	Paint doors
7:40	1921	62	12	143	Paint parts
8:10	142	19	65	682	Prime parts
9:55	236	6	91	921	Prime metal
11:05	5961	4	3	534	Paint motors

Finished last job at 11:50.

WHAT YOU ARE TO DO

Use the current date and complete the form. Compute the hours worked on each order and total hours worked. Type Actual Work Hours as follows: Work Order Number 165; 4:00–5:25 = 1 25/60 hours—type as 1:25.

EMPLOYEE NO. | SHIFT | JOB CLASS SAME AS ABOVE UNLESS SHOWN OTHERWISE | EMPLOYEE'S NAME | DATE

SHOW SHIFT → IF DIFFERENT

WORK ORDER NUMBER | REDIST. DEPT. | ACCT. | ACCT. NO. | ACTUAL WORK HRS. | JOB DESCRIPTION

START

REPORTED ON JOB TICKETS

DAILY TIME CARD

TOTAL HOURS WORKED

↑ INDICATE OVERTIME WORK WITH ✓ MARK

NAME: _______________________________ **JOB 32**

EMPLOYEE NO. SHOW SHIFT → IF DIFFERENT	SHIFT	JOB CLASS SAME AS ABOVE UNLESS SHOWN OTHERWISE	EMPLOYEE'S NAME					DATE
			WORK ORDER NUMBER	REDIST. DEPT.	ACCT.	ACCT. NO.	ACTUAL WORK HRS.	JOB DESCRIPTION
START		REPORTED ON JOB TICKETS						

DAILY TIME CARD

TOTAL HOURS WORKED

↑ INDICATE OVERTIME WORK WITH ✓ MARK

WHAT YOU SHOULD KNOW

Brown's Electric Company, 472 Commerce, Baltimore, Maryland 21222-3257, has returned one Word Processor, model 6, serial no. 320507, at $5,785; one Swivel Base Chair, model 10, serial no. 6302, at $249.95; one Workstation Desk, 66″ wide by 30″ deep, model 12, serial no. 3871, at $1,295.50.

It is necessary to have the following facts to complete the memo:

- Customer's order no. 5249, one month previous to current date.
- Requisition No. 3207.
- Date order received, 3 days following order date.
- Shipping point and date: Detroit, 3 days following date order received.
- Credit date: Current date.
- Salesperson: Donald Smith.
- Via: Railway Express.

WHAT YOU ARE TO DO

Compute the total of the Credit Memorandum and type the form. Brown's Credit Memorandum is given the number 2506.

CUSTOMER'S ORDER NO. & DATE	REQUISITION NO.	DATE ORDER REC'D	SHIPPED FROM	DATE SHIPPED	CREDIT DATE	
SALESMAN	ACCOUNT AT				VIA	

SHIPPED TO

TERMS
NET 30 DAYS

QUANTITY		MODEL	SERIAL NO.	UNIT PRICE	AMOUNT
	CREDIT MEMORANDUM				

CUSTOMER'S ORDER NO. & DATE	REQUISITION NO.	DATE ORDER REC'D	SHIPPED FROM	DATE SHIPPED	CREDIT DATE	
SALESMAN	ACCOUNT AT				VIA	

SHIPPED TO

TERMS
NET 30 DAYS

QUANTITY		MODEL	SERIAL NO.	UNIT PRICE	AMOUNT
	CREDIT MEMORANDUM				

JOB 34
Telephone Message

WHAT YOU SHOULD KNOW

You have been asked to act as receptionist in Mr. Lindsey's office and to make a memo of any messages received. Dr. John Elliott from the Board of Education called at 11:00 A.M. He asked that Mr. Lindsey call him as soon as possible. Dr. Elliott's phone number is 353-0261, Extension 271.

WHAT YOU ARE TO DO

Complete the form using the current date.

MESSAGE

Date _____________ 19 _____ Time _____________ A.M.
                                                P.M.

To ____

WHILE YOU WERE OUT

Name ____

Address____

Outside
Phone No. ____________________________ Ext. _________

Was here to see you	☐	Telephoned	☐
Called on Intercom	☐	Will call again	☐
Returned your call	☐	Please call	☐

Message ____

____

____

____

____

____

TAKEN BY

MESSAGE

Date _________________ 19 _______ Time _____________ A.M.
P.M.

To ____

WHILE YOU WERE OUT

Name ____

Address____

Outside
Phone No. _________________________________ Ext. __________

Was here to see you	☐	Telephoned	☐
Called on Intercom	☐	Will call again	☐
Returned your call	☐	Please call	☐

Message ____

____

____

____

____

____

TAKEN BY

WHAT YOU SHOULD KNOW

An order of medical supplies is being mailed to Dr. Frederick Lewis, 26 Westover Drive, Dover, Delaware 19901-6220. The package includes:

- 5 boxes tongue depressors (250 each)
- 1 gross envelopes for pills
- 1 dozen bottles—2 oz.
- 2 dozen bottles—4 oz.

No prices should be given.

WHAT YOU ARE TO DO

Fill in the form using the information indicated. Use the current date. Use your teacher's initials in space provided for "Inspected By" and your initials for "Sold By" and "Filled By." Abbreviate as follows: bx for box, gr for gross, doz for dozen; do not use periods after the abbreviations (your employer's policy).

CONTENTS—MERCHANDISE

POSTMASTER: This parcel may be opened for postal inspection if necessary. Return and Forwarding Postage Guaranteed.

DR.

SOLD BY	FILLED BY	INSPECTED BY	DATE

T

16425

CONTENTS—MERCHANDISE

POSTMASTER: This parcel may be opened for postal inspection if necessary. Return and Forwarding Postage Guaranteed.

DR.

SOLD BY	FILLED BY	INSPECTED BY	DATE

T

16425

Automobile Accident or Loss Report

WHAT YOU SHOULD KNOW

This report form is used by policyholders to report claims arising from automobile accidents or losses. Information for the report is taken from the policy and from the driver's version of the accident.

Assume that you work for Mr. Raymond K. Geiger, 230 Richmond Road, Cedar Rapids, Iowa 52402-2579, telephone 927-1364. His business address is 1927 Kevin Road, ZIP code 52401-8241, and his business telephone number is 739-2074.

While returning from school, your employer's son was involved in an accident with John Butler of 1296 Wilder Drive, Cedar Rapids, Iowa 52401-7323, telephone 736-2159. Mr. Butler was driving a 1983 Dodge (license #Iowa M-5800) west on Commerce in Cedar Rapids, Iowa, at about 4:40 P.M. Your employer's son Kenneth, age 16, was driving a blue 1982 Chevrolet (2-door hard-top, motor #M-82-1438591, license #Iowa M-5631) south on Grand Avenue. There were no traffic controls, and Kenneth saw the other car a little too late. He could not stop because the streets were wet from a misty rain. Mr. Butler told Kenneth he was sorry that he did not see him in time to stop and gave him his telephone number, 736-2159, just before Kenneth was taken to the hospital with a cut lip.

This accident occurred on Tuesday of last week, and it was reported to the police immediately. The police did not cite either driver. The estimated damage was $875 to the Geiger car and $650 to the Butler car. Both cars were damaged in the front fender area. (You determine which side of each car.) The Geiger car may be seen at 230 Richmond Road, Cedar Rapids, and the Butler car may be seen at 1269 Wilder Drive, Cedar Rapids. The other car was insured by the XYZ Company of Cedar Rapids, Policy #172. The next day Roland Woodruff of 609 Pine, ZIP Code 52403-1957, 284-7321, called Mr. Raymond Geiger to say he had seen the accident.

WHAT YOU ARE TO DO

Mr. Geiger asked you to report the claim under this policy number 42-8160-319.

AUTOMOBILE ACCIDENT OR LOSS REPORT

POLICY NO.

IMPORTANT: In the event of an accident, however slight, please answer all questions on this blank and mail it at once. In case of fatal accident or serious injury, telegraph or telephone at once.

POLICY-HOLDER

NAME	PHONE
ADDRESS	
BUSINESS ADDRESS (NAME OF FIRM, IF ANY)	BUSINESS PHONE

TIME AND PLACE

DATE AND TIME OF LOSS OR ACCIDENT	LOCATION

INSURED AUTOMOBILE

YEAR	MAKE	BODY STYLE	COLOR	IDENT. OR SERIAL NO. (THIS MUST BE SHOWN)	LICENSE NO., STATE, & YR.

OWNER OF INSURED AUTOMOBILE	ADDRESS	OTHER INSURANCE

NAME OF DRIVER	ADDRESS	AGE	PHONE

FOR WHAT PURPOSE WAS AUTOMOBILE BEING USED AT TIME OF ACCIDENT?

WAS DRIVER EMPLOYED BY YOU?	IN WHAT CAPACITY?

IF COLLISION OR COMPREHENSIVE SPECIFY DAMAGE — IF TIRE DAMAGED OR STOLEN, GIVE MILEAGE USED

WHERE MAY AUTO BE SEEN? (ADDRESS)	ESTIMATED COST OF REPAIR

IF THEFT, INDICATE (1) PROPERTY STOLEN; (2) NAME & ADDRESS OF SELLER; (3) DATE PURCHASED; PURCHASED NEW OR USED

HAVE POLICE BEEN NOTIFIED?	WHERE AND WHEN?	DID POLICE INVESTIGATE?

OFFICE USE ONLY
NATB ☐ TWX ☐ 200

DAMAGE TO PROPERTY OF OTHERS

OWNER	ADDRESS	PHONE
OTHER DRIVER	ADDRESS	PHONE

LIST DAMAGE	ESTIMATED COST OF REPAIR
IF AUTOMOBILE, SHOW MAKE AND YEAR MODEL	LICENSE NO. AND STATE

WHERE MAY DAMAGED PROPERTY BE SEEN? (ADDRESS)

WAS OTHER CAR INSURED?	NAME AND ADDRESS OF COMPANY, AND POLICY NUMBER

PERSONS INJURED

NAME	ADDRESS	(CHECK ONE)			EXTENT OF INJURIES
		PASSENGER INSUR-ED'S CAR	OTHER CAR	PEDES-TRIAN	

WITNESSES

NAME	ADDRESS	PHONE

DATE OF THIS REPORT

.... 19...... SIGNATURE .... INSURED OR DRIVER

NAME: _________________________ JOB 36

213

HOW DID ACCIDENT OR LOSS OCCUR?

DESCRIPTION OF ACCIDENT OR LOSS

Tell in your own words just what happened.

Also furnish any other information you consider important.

....
....
....
....
....
....
....
....
....
....

WERE YOU AT FAULT?

WHAT STATEMENT DID OTHER PARTY MAKE AS TO ACCIDENT?

CONDITION OF PAVEMENT | WEATHER CONDITIONS

PLEASE COMPLETE THIS DIAGRAM. SHOW NAMES OF STREETS, DIRECTION AND POSITION OF AUTOMOBILES, AND POINT OF CONTACT.

Indicate by arrow direction of north   N ———►

Instructions:

(1) Use solid line to show path of vehicles before accident ———►
(2) Use dotted line to show path of vehicles after accident ·········►
(3) Number each vehicle and show direction of travel ———►
(4) Show pedestrian by ———► O

AUTOMOBILE ACCIDENT OR LOSS REPORT

POLICY NO.

IMPORTANT: In the event of an accident, however slight, please answer all questions on this blank and mail it at once. In case of fatal accident or serious injury, telegraph or telephone at once.

POLICY-HOLDER

NAME | PHONE

ADDRESS

BUSINESS ADDRESS (NAME OF FIRM, IF ANY) | BUSINESS PHONE

TIME AND PLACE

DATE AND TIME OF LOSS OR ACCIDENT | LOCATION

INSURED AUTOMOBILE

YEAR | MAKE | BODY STYLE | COLOR | IDENT. OR SERIAL NO. (THIS MUST BE SHOWN) | LICENSE NO., STATE, & YR.

OWNER OF INSURED AUTOMOBILE | ADDRESS | OTHER INSURANCE

NAME OF DRIVER | ADDRESS | AGE | PHONE

FOR WHAT PURPOSE WAS AUTOMOBILE BEING USED AT TIME OF ACCIDENT?

WAS DRIVER EMPLOYED BY YOU? | IN WHAT CAPACITY?

IF COLLISION OR COMPREHENSIVE SPECIFY DAMAGE — IF TIRE DAMAGED OR STOLEN, GIVE MILEAGE USED

WHERE MAY AUTO BE SEEN? (ADDRESS) | ESTIMATED COST OF REPAIR

IF THEFT, INDICATE (1) PROPERTY STOLEN; (2) NAME & ADDRESS OF SELLER; (3) DATE PURCHASED; PURCHASED NEW OR USED

HAVE POLICE BEEN NOTIFIED? | WHERE AND WHEN? | DID POLICE INVESTIGATE?

OFFICE USE ONLY
NATB ☐ TWX 200

DAMAGE TO PROPERTY OF OTHERS

OWNER | ADDRESS | PHONE

OTHER DRIVER | ADDRESS | PHONE

LIST DAMAGE | ESTIMATED COST OF REPAIR

IF AUTOMOBILE, SHOW MAKE AND YEAR MODEL | LICENSE NO. AND STATE

WHERE MAY DAMAGED PROPERTY BE SEEN? (ADDRESS)

WAS OTHER CAR INSURED? | NAME AND ADDRESS OF COMPANY, AND POLICY NUMBER

PERSONS INJURED

NAME	ADDRESS	(CHECK ONE)			EXTENT OF INJURIES
		PASSENGER INSUR-ED'S CAR	OTHER CAR	PEDES-TRIAN	

WITNESSES

NAME	ADDRESS	PHONE

DATE OF THIS REPORT

19...... | SIGNATURE .... INSURED OR DRIVER

NAME: _________________________ JOB 36

215

HOW DID ACCIDENT OR LOSS OCCUR?

DESCRIPTION OF ACCIDENT OR LOSS

Tell in your own words just what happened.

Also furnish any other information you consider important.

WERE YOU AT FAULT?

WHAT STATEMENT DID OTHER PARTY MAKE AS TO ACCIDENT?

CONDITION OF PAVEMENT | WEATHER CONDITIONS

PLEASE COMPLETE THIS DIAGRAM. SHOW NAMES OF STREETS, DIRECTION AND POSITION OF AUTOMOBILES, AND POINT OF CONTACT.

Indicate by arrow N
direction of north

Instructions:
(1) Use solid line to show path of vehicles before accident
(2) Use dotted line to show path of vehicles after accident
(3) Number each vehicle and show direction of travel
(4) Show pedestrian by

Records Transmittal Advice and Receipt

WHAT YOU SHOULD KNOW

As soon as possible after the close of the business year, records no longer needed in all offices of XYZ Company are transferred to the Central Office building for microfilming. Anne Wells, Payroll Office, transmits the following records to the General Records File. Use the current date and General Records Register #2692.

CARTON #	RECORD	CONSISTING OF	FILE REFERENCE
46	Accounts Receivable	A–Z	A1
47	Sales Tax Forms A12	Jan.–Dec.	A2
48	General Correspondence	A–Z	A3
48	Sales Correspondence	A–Z	A4
49	Cancelled Checks	1001–1075	A5
49	Contracts	150–260	A6
49	Telephone Messages	Jan.–Dec.	A7

All records are from last year.

WHAT YOU ARE TO DO

Fill in the blanks with the information provided; do not abbreviate the months. Blanks at the bottom of the page will be filled in by the General Records Section.

General Records
Register No.___________

RECORDS TRANSMITTAL ADVICE AND RECEIPT

TO: <u>GENERAL RECORDS SECTION</u>   DATE:___________________________

ATTN:___________________________   FROM:___________________________

OFFICE:_________________________   OFFICE:___________________________

We are forwarding the following described records for retention in General Records
Files. Attached hereto is the original bill of lading showing the delivering
carrier and the date shipped.

Carton No	Year	Record	Consisting of	File Reference

ACKNOWLEDGMENT OF RECEIPT

_____ Receipt of Shipment - Date ____________   General Records

_____ Filing Completed - Date ______________   By_________________________
                                                  Supervisor

RECORDS TRANSMITTAL ADVICE AND RECEIPT

TO:_GENERAL RECORDS SECTION__   DATE:_______________________________

ATTN:___________________________   FROM:______________________________

OFFICE:_________________________   OFFICE:____________________________

We are forwarding the following described records for retention in General Records
Files. Attached hereto is the original bill of lading showing the delivering
carrier and the date shipped.

Carton No	Year	Record	Consisting of	File Reference

ACKNOWLEDGMENT OF RECEIPT

_____ Receipt of Shipment - Date ____________   General Records

_____ Filing Completed - Date ______________   By____________________________
                                                        Supervisor

JOB 38
Back-order or Partial Material Received Report

- -

WHAT YOU SHOULD KNOW

The Public Relations Department has received six of nine items ordered one month ago today from Atop Paper Company, 75 Fulton, Boston, Massachusetts 02109-4836. The remaining items will have to be back-ordered.

WHAT YOU ARE TO DO

Fill in the blanks using the information given above. The original Purchase Order No. was 604-142. Use the following back order:

2 gross	#2 pencils
7 quire	Mimeograph stencils
1 quire	16# bond, 8½ × 11 in.

Omit blanks at the bottom of the page.

BACK ORDER OR
PARTIAL MATERIAL RECEIVED REPORT

ORIGINAL PURCHASE ORDER NO.

____
UNIT ISSUING THIS FORM

☐ THIS FORM USED AS "BACK ORDER"

☐ THIS FORM USED AS "PARTIAL MATERIAL RECEIVED REPORT"

DATE OF
ORDER________________ ____19____

____
SUPPLIER

DATE OF
THIS FORM________________________19____

____
ADDRESS

PLACE CHARGE AT BOTTOM OF THIS PAGE

QUANTITY		DESCRIPTION	DO NOT USE
RECEIVED	BACK ORDERED		

PLEASE FURNISH ALL INFORMATION REQUESTED—SEE INSTRUCTIONS ON REVERSE SIDE

To Be Used For

Charge

Name of Shipper

Shipped Via ⟶ ☐ Express ☐ Truck Line ☐ Parcel Post ☐ Supplier's Delivery ☐ Our Pickup ☐ Rail Freight

Car Initial and Number

Name of Transp. Co.

Trans. Bill No.   Date

☐ Prepaid ☐ Coll.

Amt. Charges Paid At
Destination $__________

Transp. Pay. Reference

Date Received

Vendor Packing List or Delivery Slip No.

Entered in Voucher No.

Received By

Approved

BACK ORDER OR
PARTIAL MATERIAL RECEIVED REPORT

ORIGINAL PURCHASE ORDER NO.

____
UNIT ISSUING THIS FORM

☐ THIS FORM USED AS "BACK ORDER"

☐ THIS FORM USED AS "PARTIAL MATERIAL RECEIVED REPORT"

DATE OF ORDER_______________ _____19_____

____
SUPPLIER

DATE OF THIS FORM_______________19_____

____
ADDRESS

PLACE CHARGE AT BOTTOM OF THIS PAGE

QUANTITY		DESCRIPTION	DO NOT USE
RECEIVED	BACK ORDERED		

PLEASE FURNISH ALL INFORMATION REQUESTED—SEE INSTRUCTIONS ON REVERSE SIDE

To Be Used For

Charge

Name of Shipper

Shipped Via ⟶ ☐ Express ☐ Truck Line ☐ Parcel Post ☐ Supplier's Delivery ☐ Our Pickup ☐ Rail Freight

Car Initial and Number

Name of Transp. Co.

Trans. Bill No.   Date

☐ Prepaid ☐ Coll.

Amt. Charges Paid At Destination $_______

Transp. Pay. Reference

Date Received

Vendor Packing List or Delivery Slip No.

Entered in Voucher No.

Received By

Approved

NAME: _______________________   JOB 38   227

Accident Investigation Report

WHAT YOU SHOULD KNOW

At 9:15 A.M. last Monday, Darryl Tripp, age 42, a tool and die maker on the first shift at the Midwest Branch of Robots, Inc., was taken to the emergency room of the local hospital to have several metal shavings removed from his right eye.

Mr. Tripp, who has been employed by the company for 12 years, was wearing a new style of safety glasses at the time of the accident. He said that a draft caused by an outer door being opened blew the shavings up under his glasses.

This is the third accident of this type in the last month since issuing the new style of glasses to our employees; therefore, the use of this style of glasses (equipment) has been discontinued. The old style of safety glasses, which fit against the face, have been reissued to the employees. This action should cause a decrease in the number of eye-related injuries.

WHAT YOU ARE TO DO

Type the Accident Investigation Report using the information provided.

ACCIDENT INVESTIGATION REPORT

Name		Age	Date & Time of Accident
Branch	Department - Shift	Job	How long on this job?

NATURE OF INJURY?

> Area of body injured as
> a result of accident.

WHAT HAPPENED?

> Describe what took place
> or what caused you to
> make this investigation.

WHY DID IT HAPPEN?

> Get all the facts by studying
> the job and
> situation involved.
> Question by use of WHY
> WHAT - WHERE - WHEN
> WHO - HOW

WHAT SHOULD BE DONE?

> Determine which of the
> 12 items under EMP
> require attention.

Equipment	Material	People
Select	Select	Select
Arrange	Place	Place
Use	Handle	Train
Maintain	Process	Lead

WHAT HAVE YOU DONE THUS FAR?

> Take or recommend action,
> depending on your authority.
> Follow up
> was action effective?

HOW WILL THIS IMPROVE OPERATIONS?

> OBJECTIVE
> ELIMINATE JOB HINDRANCES

NAME: _______________________________   JOB 39

ACCIDENT INVESTIGATION REPORT

Name		Age	Date & Time of Accident
Branch	Department - Shift	Job	How long on this job?

NATURE OF INJURY?

> Area of body injured as
> a result of accident.

WHAT HAPPENED?

> Describe what took place
> or what caused you to
> make this investigation.

WHY DID IT HAPPEN?

> Get all the facts by studying
> the job and
> situation involved.
> Question by use of WHY
> WHAT - WHERE - WHEN
> WHO - HOW

WHAT SHOULD BE DONE?

> Determine which of the
> 12 items under EMP
> require attention.

Equipment	Material	People
Select	Select	Select
Arrange	Place	Place
Use	Handle	Train
Maintain	Process	Lead

WHAT HAVE YOU DONE THUS FAR?

> Take or recommend action,
> depending on your authority.
> Follow up
> was action effective?

HOW WILL THIS IMPROVE OPERATIONS?

> OBJECTIVE
> ELIMINATE JOB HINDRANCES

Supplies Requisition

WHAT YOU SHOULD KNOW

The Accounting Department at Newark, New Jersey, wishes the following supplies delivered to Mr. Robert Burnside in Room 201A:

30 pads	7202L Columnar Pads
1 doz.	24120L Post Binder
12 sets	50156L Post Binder Indexes
4 doz.	420L Ledger Sheets

WHAT YOU ARE TO DO

Fill in all blanks that would normally be typed. Use the current date. Remember to type neatly in relation to the line.

INSTRUCTIONS

A Separate Requisition should be prepared and sent in a Large Intra Company Envelope to the room as indicated, for Each of the following items.

SUPPLIES (Pencils, etc.) & ENVELOPES	1653	* MACHINES & EQUIPMENT	1637
FORMS	1643	* FURNITURE	1016
POSTAGE	149-M		

FOR DEPARTMENT	DELIVER TO	ROOM NO.	DATE

QUANTITY	FORM OR SUPPLY NO.	DESCRIPTION	DO NOT USE THESE COLUMNS		
			LOCATION	DATE SHIPPED	BACK ORDERED DO NOT REORDER

SIGNATURE	SIGNATURE	REQUISITIONED BY	* REQUIRES DEPT. SUPERVISORS APPROVAL

NAME: _______________________   JOB 40

237

DO NOT FOLD
(SEND IN ENVELOPE)

REQUISITION

INSTRUCTIONS

A Separate Requisition should be prepared and sent in a Large Intra Company Envelope to the room as indicated, for Each of the following items.

SUPPLIES (Pencils, etc.) & ENVELOPES	1653	* MACHINES & EQUIPMENT	1637
FORMS	1643	* FURNITURE	1016
POSTAGE	149-M		

FOR DEPARTMENT	DELIVER TO	ROOM NO.	DATE

QUANTITY	FORM OR SUPPLY NO.	DESCRIPTION	DO NOT USE THESE COLUMNS		
			LOCATION	DATE SHIPPED	BACK ORDERED DO NOT REORDER

SIGNATURE	SIGNATURE	REQUISITIONED BY	* REQUIRES DEPT. SUPERVISORS APPROVAL

Index of Job Classifications